SCOTTISH STEAM
in the 1950s and '60s

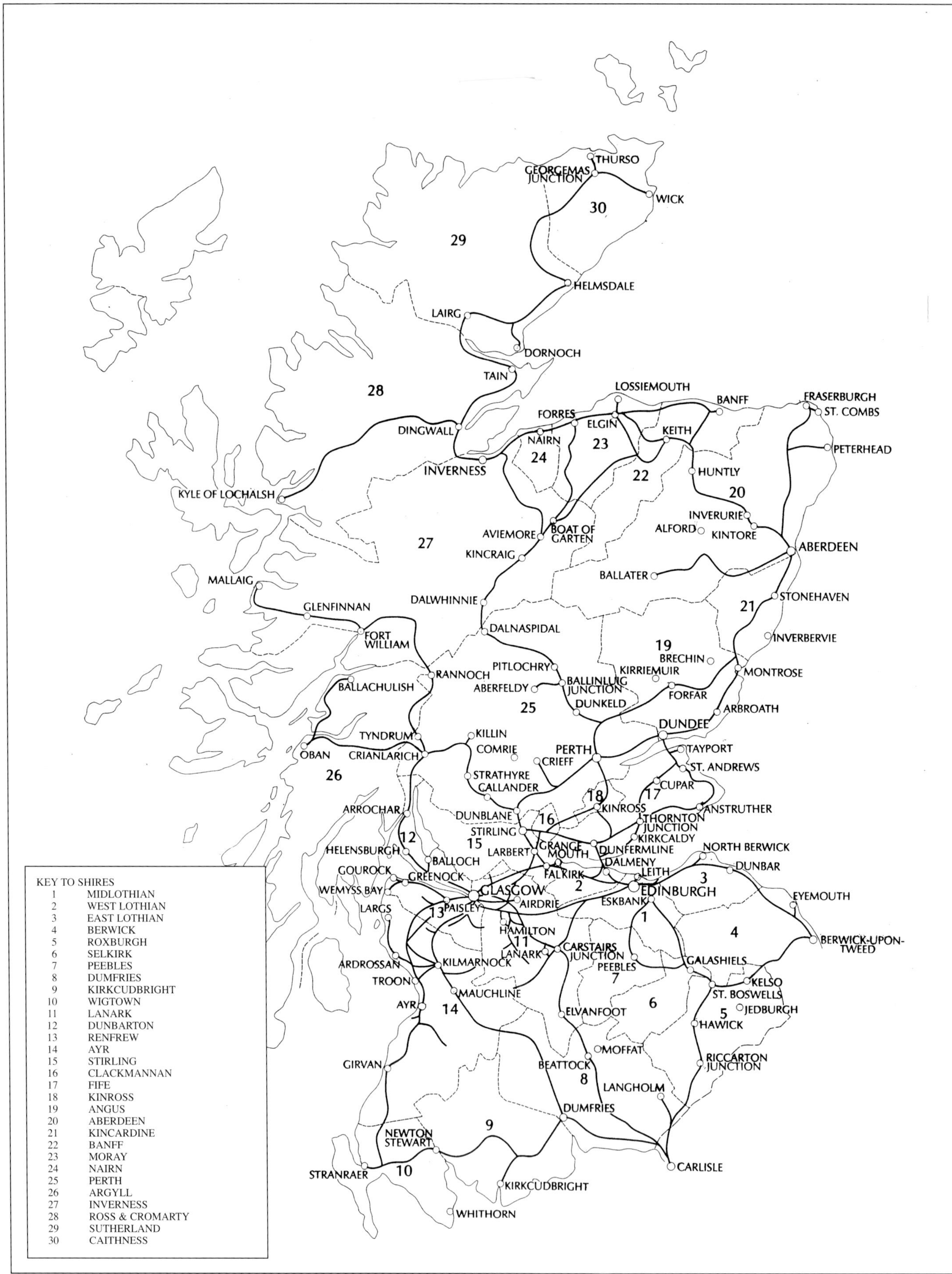

Railway map of Scotland showing extent of the passenger network c1960

SCOTTISH STEAM
in the 1950s and '60s

DAVID ANDERSON

OPC

Oxford Publishing Co.

Class V3 2-6-2T No. 67606 of Edinburgh St Margaret's depot comes off the Forth Bridge and into Dalmeny station with an all-stations local passenger train from Dunfermline to Edinburgh Waverley on 15th June 1957.

Half title page
Steam power personified – an all-out effort from Class 8P 'Princess Royal' 4-6-2 No. 46212 *Duchess of Kent* as she tackles the last mile of Beattock Bank with the 11.5 am Birmingham New Street to Glasgow Central express in July 1959.

A catalogue record for this book is available from the British Library.

ISBN 0 86093 479 9

Oxford Publishing Co. is an imprint of Haynes Publishing, Sparkford, Nr Yeovil, Somerset BA22 7JJ

Tel: 01963 440635   Fax: 01963 440001
Int. tel: +44 1963 440635   Int. fax: +44 1963 440001
E-mail: sales@haynes-manuals.co.uk
Web site: http:/www.haynes.com

Printed in Hong Kong

Typeset in Times Roman Medium

As part of our ongoing market research, we are always pleased to receive comments about our books, suggestions for new titles, or requests for catalogues. Please write to: The Editorial Director, Oxford Publishing Co., Sparkford, Near Yeovil, Somerset, BA22 7JJ

Contents

Introduction

During the early 1950s, when steam locomotive power still reigned supreme in Scotland, the prospect of the wholesale slaughter of branch lines and the complete elimination of steam traction could hardly have been contemplated.

As passenger and rail freight traffic steadily dwindled in the face of road competition, the inevitable ensued and the threatened railway closures became reality. The social and economic character of large areas of Scotland was irretrievably altered as lines and stations were abandoned and many parts of the country, especially the Borders and the North East, lost their life lines and became empty spaces on the railway map. Fortunately, due to the dedication and foresight of enthusiasts and historians, much of this heritage has been recorded in word and picture for posterity.

This book is not intended to be a technical or historical account of mid-century Scottish Steam, but rather a representative selection of photographs from the collections of David Anderson, Stuart Sellar and Ian Swanson. It is hoped that it will help to evoke nostalgia and recall memories of the railway system north of the border in the 1950s and early 1960s, at a time when such everyday scenes were taken so much for granted in the glorious days of steam.

Pages 7 to 120 of the book contain illustrations geographically based within the former Scottish shires which existed prior to the boundary reorganisation of 1975. Pages 125 to 138 are devoted to a photographic survey of all locomotive class types, with minor exceptions as noted, which were in regular service and officially allocated to Scottish Region depots in 1954 followed by a miscellany of railway related subjects including Royal Trains and more recent pictures of the early days of preserved steam at work in Scotland.

Our railway journey begins in the capital city of Edinburgh and covers some of the former Caledonian Railway and North British Railway strongholds of steam in the Lothians, thence south to the rolling hills of the Border country, west through Dumfriesshire and the South West of Scotland to Stranraer before heading north once again to the grime of industrial Lanarkshire. In contrast, we linger by the lineside on the notorious moorland climb of Beattock Bank over the Southern Uplands on the West Coast Main Line.

The delightful and photogenic 'push-and-pull' steam service to Arrochar is illustrated, and we then retrace our steps in a southern direction and glimpse some of the Clyde Coast stations and the railway diversities of Ayrshire. Travelling eastwards from Stirling in the Central Lowlands, we head through the Kingdom of Fife with its variety of railway interest, and across the Firth of Tay – the scene of Sir Thomas Bouch's ill-fated bridge collapse – to Dundee in Angus. The former Great North of Scotland Railway territory north of Aberdeen is visited, including a brief pause at Keith, before proceeding by single-track line to the ever-popular coastal termini of Oban, Mallaig and the Kyle of Lochalsh. Our journey concludes at Wick and Thurso after travelling through the scenic grandeur of the Highlands over wide and windswept tracts of land to the northern outposts of Scottish steam working.

It is impossible in a volume of this size to do justice to the infinite variety and wealth of railway interest to be found in Scotland during this period. Inevitably, some gaps do exist in the pictorial coverage, but it is hoped that the reader will understand if a particular location, favourite steam locomotive at work, or a particular stretch of line has been omitted. No photographic album can be produced without the help of others, and I am indebted to the two contemporary photographers, who have kindly made available for selection their precious collection of negatives.

Finally, grateful thanks are due to the many railway servants and staff throughout the Scottish Region of British Railways who usually managed to turn a blind eye to our frequent wanderings!

I am especially indebted to the shed and locomotive staff, many of whom failed, understandably, to share our enthusiasm for what was to them often dangerous and dirty work. If some of the atmosphere of the Scottish railway scene has been conveyed to the reader in the following pages, the long hours spent with the camera by the lineside and at the steam shed, attempting to overcome the whims and vagaries of the Scottish climate, will have been worthwhile.

David Anderson

Midlothian, West Lothian, East Lothian

The railway gateway to Scotland, Edinburgh Waverley, the former North British Railway and London & North Eastern Railway station at the eastern end of Princes Street, is situated in the valley between the Old and New Towns of Scotland's capital city. Looking west from above the Mound tunnel on the approaches to the station, Class A3 4-6-2 No. 60073 *St Gatien* drifts through Princes Street Gardens below the castle ramparts, to take charge of a morning southbound East Coast Main Line express from Edinburgh Waverley station on 8th September 1957, a scene since altered by modernisation.

Waverley station, Edinburgh

A general view of the west end of the station, its 19 platforms covering a 23-acre site dominated by the Waverley Bridge and the elegant North British Railway Hotel, now renamed The Balmoral.

British Railways Standard Class 7MT 'Britannia' 4-6-2 No. 70020 *Mercury*, previously allocated to the Western Region, enters the central platform as a Class 4 English Electric diesel locomotive leaves the station with an express passenger train to Dundee and Aberdeen during the final months of steam working in Scotland.

Edinburgh Waverley station 'pilots'

Well-turned-out Edinburgh Waverley station east end 'pilots', former North British Railway Class J83 0-6-0Ts Nos 68470 and 68477 of Edinburgh St Margaret's depot, pause between shunting duties beneath the walls of the Old Calton gaol, on a hot summer's day in 1956.

Edinburgh Haymarket depot

A familiar scene at the west end of Haymarket depot in steam days, on 14th April 1958, from left to right, Class D30/2 4-4-0 No. 62418 *The Pirate* of Thornton Junction shed, Class A2 4-6-2 No. 60536 *Trimbush* of Haymarket depot, and Class A3 4-6-2 No. 60079 *Bayardo*, one of the four Carlisle Canal-based Gresley Pacifics, are serviced in the shadow of the locomotive coaling plant, a once-dominant feature of the area.

'Top-link' Pacific

With its train nameboard reversed until coupling up to the southbound 'non-stop' 'Elizabethan' express to London King's Cross, Gresley Class A4 4-6-2 No. 60011 *Empire of India* gleams in the morning sun as she slowly heads past Haymarket Central Junction signal box between the depot and Haymarket station en route to Edinburgh Waverley on 26th August 1956. One of Haymarket's 'top-link' Pacifics from its 1954 allocation of seven Class A4s, No. 60011 built at Doncaster Works as No. 4490 in 1937, survived in main line service until May 1964.

Built for the LNER at Doncaster Works in July 1934, an immaculate Gresley Class A3 4-6-2, No. 60035 *Windsor Lad*, fitted with double chimney, prepares to leave its home depot at Edinburgh Haymarket to work a southbound Edinburgh Waverley to Newcastle express in August 1958. Numerically, the first engine of the class and one of an allocation of 15 A3 Pacifics based at Haymarket in 1954, No. 60035 was withdrawn from British Railways' service in September 1961.

Edinburgh Haymarket shed Pacifics

Maximum effort from Class A2 4-6-2 No. 60534 *Irish Elegance* as she powers towards the camera with the 9.10 am Leeds City to Glasgow Queen Street express – 'The North Briton' – under the overbridge carrying the former Caledonian Railway's Leith North to Edinburgh Princes Street branch line at Haymarket Central Junction, on 21st April 1956.

The eastern frontage of Haymarket station, the original 1842 terminus of the Edinburgh & Glasgow Railway. The building still retains its large blue and white painted wooden nameboard flanked by British Railways totem signs. These embellishments were removed when the station underwent modernisation in the 1970s.

Situated a mile to the west of Edinburgh's city centre, the listed building is a classic design in stone with an entrance surrounded by columns and a hooded clock at roof level.

Haymarket station, Edinburgh

Class B1 4-6-0 No. 61244 *Strang Steel*, one of Edinburgh Haymarket's 1954 allocation of eight engines of the class, pauses briefly at Haymarket station whilst running 'light engine', before heading tender first, through the gloom of the 1,009 yards-long Haymarket Tunnel and on towards Princes Street Gardens and Edinburgh Waverley station to work a morning semi-fast passenger train to Dundee Tay Bridge in the summer of 1958.

St Margaret's locomotive depot, situated on the eastern approaches to the city, was bisected by the East Coast Main Line. The main running shed was located on the down side of the line whilst on the up side, an open roundhouse handled the storage and repair of the smaller ex-North British Railway tank engine types. The depot provided steam power for freight workings, dockyard shunting and engines for local passenger trains in the Lothian, Fife and Border areas. St Margaret's, together with its main sub-depots at Seafield and South Leith and five smaller sheds at Dunbar, Galashiels, Granton, North Berwick and Peebles, had an allocation of over 220 engines in 1954, the largest in Scotland. To the relief of local residents and railway staff alike, the cramped and dirty depot was closed in April 1967 and the site now accommodates office blocks and the Meadowbank Sports Stadium. At the shed roundhouse under repair on 14th June 1959 were, from left to right, Class J88 0-6-0T No. 68338, Class J83 0-6-0T No. 68477 and Class N15/1 0-6-2T No. 69173.

Edinburgh St Margaret's depot

The 27-ton weight of ex-North British Railway Class Y9 0-4-0ST No. 68118 is hoisted up for mechanical inspection at the open roundhouse at St Margaret's on 14th August 1955.

St Leonards branch, Edinburgh

Opened in 1831 and originally horse-drawn, the old Edinburgh & Dalkeith Railway ran a short branch line from Duddingston to St Leonards to transport supplies of coal to the city, and later, to carry passengers.

In the shadow of Arthur's Seat, Edinburgh, Class J35 0-6-0 No. 64535 attacks the 1 in 30 incline towards the rock tunnel and St Leonards terminus with a lightweight goods train on a summer's morning in 1956. The 'Innocent Railway' was closed on 5th August 1968 and the former trackbed is now in use as a city walkway.

Edinburgh suburban line passenger

Connecting the southern environs of the city by rail from Waverley station, the Edinburgh circle line served nine suburban destinations. Edinburgh St Margaret's shed's clean Class V3 2-6-2T No. 67624 approaches Craiglockhart, between Gorgie East and Morningside Road stations with an afternoon eastbound passenger train partly composed of Gresley ex-LNER articulated coaching stock in August 1955. The suburban line was closed to passengers on 10th September 1962, but remains open as a freight traffic link bypassing the city centre. No. 67624 was withdrawn from service in September 1960.

Ivatt Class 4MT 2-6-0 No. 43138 of Bathgate shed, leaves a smoke screen behind as she tackles the gradient between Gorgie East and Craiglockhart stations on the Edinburgh suburban line with an eastbound train of washed coal from Grangemouth, possibly destined for Portobello Power Station, on 6th April 1957.

Edinburgh suburban line mineral traffic

At the same location, on 29th September 1956, Class 2F 0-6-0 No. 57287 of Grangemouth shed, in remarkably good external condition, passes under the Edinburgh Princes Street to Glasgow Central main line and starts the climb from Gorgie East station with a train of coal empties from Grangemouth to Leith South yards, to the east of Edinburgh.

Opened in June 1962, Millerhill yard to the east of the city, was a concentration and marshalling point for Scottish Region freight workings, replacing the yards at Portobello, South Leith and Niddrie. A clean, green-liveried Class V2 2-6-2, No. 60931, leaves the down yard with the 6.27 pm freight to Craiginches (Aberdeen) on 25th July 1963.

Millerhill yards

British Railways Standard Class 7MT 'Britannia' 4-6-2 No. 70020 *Mercury* heads for Carlisle via the Waverley route with the 3.25 pm freight from Millerhill on 8th May 1963. A Class J38 0-6-0 heading a local mineral working occupies the adjoining spur line from Monktonhall Junction.

The northern end of the Waverley route, from Edinburgh to Carlisle via Melrose and Hawick, diverged from the East Coast Main Line at Portobello East Junction. With the climb to Fala Summit ahead, Class A2/3 4-6-2 No. 60519 *Honeyway* of Edinburgh Haymarket depot, passes Niddrie North Junction with an afternoon passenger train from Edinburgh Waverley to Carlisle on 17th October 1955.

Portobello and Musselburgh

The mile-long Musselburgh branch opened in July 1847 from the East Coast Main Line at Newhailes Junction, between Joppa and Inveresk. British Railways Standard Class 2MT 2-6-0 No. 78048 is shown leaving the town with a local passenger train to Edinburgh Waverley in the late 1950s. Closure to passenger traffic took place in September 1964 and the former trackbed at this point is now a roadway. A new station, on the main line, and ¾ mile from Musselburgh town centre, was opened in 1988 and provides a commuter service to Edinburgh Waverley.

Reid designed and built at Cowlairs Works, Glasgow in 1913, *Glen Douglas* became LNER No. 9256 in 1924 and No. 2469 in 1946. Restored to North British Railway livery whilst still in traffic, British Railways No. 62469 became No. 256 and was used for working special trains and enthusiasts' rail tours before permanent retirement as a static exhibit at the Museum of Transport, Glasgow. A rail tour hauled by the preserved 4-4-0 travelled over disused lines to the south east of Edinburgh on 29th August 1959 and the special train is seen near Gilmerton on the former North British branch line to Glencorse, Midlothian. The engine has now been transferred to the Bo'ness & Kinneil Railway and may return in the future to its former haunts to work passenger excursions over the West Highland line.

The preserved 'Glen' class 4-4-0

The driving wheel splasher of No. 256 showing the attractive style of shaded lettering and the Cowlairs Works brass builder's plate.

At the western end of Edinburgh's Princes Street, the Caledonian Railway built its seven-platform terminus in 1894. Known locally as the 'Caley', the station in its heyday ran through express trains to London via Carlisle, Birmingham, Glasgow Central, Stirling, Perth, Oban and local services to the suburban destinations of Leith, Barnton and Balerno. After closure of the station in September 1965, its remaining passenger services were transferred to Edinburgh Waverley via the Duff Street connection near Haymarket station.

On a spring morning in 1956, a young enthusiast admires Class 8P 'Duchess' 4-6-2 No. 46220 *Coronation*, the first engine of the class, as she leaves the terminus with the lightweight 11.37 am all-stations passenger train to Glasgow Central, a regular 'filling-in' working for the Polmadie-based 'Pacifics' and 'Royal Scot' class 4-6-0s.

Princes Street station, Edinburgh

Princes Street station, Edinburgh as it looked not long after closure in 1965. The station building which adjoined the Caledonian Hotel has been demolished and the derelict area is at present in use as a city centre car park.

A few minutes after departure from Edinburgh Princes Street station, Class 6P 'Jubilee' 4-6-0 No. 45691 *Orion* of Glasgow Polmadie depot, and recently ex-works, passes Dalry Road depot with a late afternoon stopping passenger train to Glasgow Central via Midcalder and Shotts in July 1957. The former trackbed of the railway at this point now forms part of the city's Western Approach road system.

Edinburgh Princes Street station departures

Glasgow Polmadie-based Class 7P 'Royal Scot' 4-6-0 No. 46121 *Highland Light Infantry, The City of Glasgow Regiment* makes light work of the three-coach all-stations 11.37 am Edinburgh Princes Street to Glasgow Central passenger train; a photograph taken near Balerno Junction on the city's western outskirts in July 1956.

In rural surroundings only a mile from the Edinburgh city centre, Class 2P 0-4-4T No. 55210 was in charge of an afternoon local branch passenger train when photographed on 17th May 1955 between Craigleith and Murrayfield on the Leith North to Edinburgh Princes Street branch line. Opened in 1879, the Leith branch also served the former Caledonian Railway stations at Newhaven, Granton Road and Dalry Road, Edinburgh. A halt at East Pilton was opened by the LMSR in December 1934. Passenger services on the Leith branch were discontinued from 30th April 1962.

Leith branch

McIntosh ex-Caledonian Railway Class 2P 0-4-4T No. 55124 of 1895 vintage and the last remaining 9 class with railed coal bunker, waits at Leith North terminus with the 'Pentland–Tinto Express' rail tour, a special train which was run to Broughton, Peebles-shire via Midcalder, Carstairs Junction and Symington on 30th September 1961.

Class 2F 0-6 0T No. 47163 leaves the Edinburgh Princes Street to Glasgow Central main line at Balerno Junction, and heads towards Colinton and Balerno with the daily branch goods train, on a spring day in the mid-1950s. More of a loop line than a branch, the railway served stations at Colinton, Juniper Green, Currie and Balerno and rejoined the Glasgow and Carstairs route at Ravelrig Junction. Apart from this regular working, it was unusual to see any of the five Scottish Region based 'Dock Tanks' in action on the main line, their activities being spent on more restricted duties in and around the docks at Granton and Greenock.

Balerno branch

On the same working, No. 47163 of Dalry Road shed, Edinburgh, climbs up the winding valley of the Water of Leith between Colinton and Juniper Green, Midlothian, with the branch goods train. Originally serving paper mills, a tannery, two stone quarries, and the sidings of a company famous for its porridge oats, the former Caledonian Railway line was closed to passengers in October 1943 and goods traffic was discontinued in December 1967.

Extensive sidings existed around the River Forth port of Granton, Midlothian, for supplying coal to Edinburgh Gas Works and to serve the busy dock areas. Class N15/1 0-6-2T No. 69187 of Dalry Road shed shunts coal wagons in the yards at Granton on 16th May 1956.

Edinburgh Dalry Road shed locomotive workings

A wave from the letter sorters to the photographer as, hurrying to connect with the southbound 'West Coast Postal' at Carstairs Junction, Class 4MT 2-6-4T No. 42173 speeds through Kingsknowe on the city's western outskirts, with the Edinburgh mail train from Princes Street station on 5th July 1955.

In the summer months during the mid-1950s and early 1960s, a through express passenger train service ran to Edinburgh Princes Street from Birmingham New Street, the train usually being worked by a Crewe North-based 4-6-0. At Edinburgh Dalry Road shed, in June 1957, Class 5MT 4-6-0 No. 45434 of Crewe North shed is being fuelled in readiness for the following day's return working. Class 5MT 4-6-0 No. 44811, a visitor from Leicester Midland shed, waits its turn to move forward to the hand-operated coaling stage.

Midland Region visitors

An unrebuilt Class 6P 'Patriot' 4-6-0, No. 45503 *The Royal Leicestershire Regiment* of Crewe North shed, a class of engine rarely seen in the capital city, stands over the ash-pits at Dalry Road depot, Edinburgh after working through from Birmingham New Street to Edinburgh Princes Street in July 1959. In the background, an ex-Caledonian Railway 0-6-0 shunts in the shed yards. Originally named *The Leicestershire Regiment* the locomotive's nameplates were recast to incorporate the *Royal* title in 1948.

Recently returned to main line service after a general overhaul at Darlington Works, Class D49/1 'Shire' 4-4-0 No. 62718 *Kinross-shire* comes off the Forth Bridge and into Dalmeny station with a Thornton Junction to Edinburgh Waverley passenger train on 23rd July 1957.

Dalmeny, West Lothian

A general view of Dalmeny station looking north on the same day with Class D11/2 'Director' 4-4-0 No. 62691 *Laird of Balmawhapple* leaving on a Dunfermline to Edinburgh Waverley stopping passenger train whilst Class K2/1 2-6-0 No. 61721 waits in an adjoining siding.

With the spans of the Forth Bridge just visible in the background, No. 60528 *Tudor Minstrel*, one of two Class A2 Pacifics based at Dundee Tay Bridge shed in 1954, charges out of Dalmeny Junction station with the 4 pm Aberdeen to Edinburgh Waverley express on 23rd May 1959.

Newcastle Heaton depot-based Class V2 2-6-2 No. 60976, hauling a southbound freight train, waits on the up slow line for Class B1 4-6-0 No. 61260 to pass with a Dundee to Edinburgh Waverley semi-fast passenger train working before gaining access to the main line at Dalmeny Junction, on 3rd May 1959.

On its regular duty with the daily 'Ferry Goods' from South Queensferry to Haymarket, Edinburgh, Class J36 0-6-0 No. 65243 *Maude* trundles along the Aberdeen to Edinburgh Waverley main line near Dalmeny Junction on 8th June 1957. *Maude*, a Neilson & Company of Glasgow-built veteran of 1891 with overseas service during World War I, has been preserved and restored to North British Railway goods livery. The engine is maintained at the Scottish Railway Preservation Society depot at Bo'ness.

Dalmeny Junction

Class A3 4-6-2 No. 60043 *Brown Jack* of Edinburgh Haymarket depot slows for the Forth Bridge crossing on the approaches to Dalmeny station with a 'Royal Canadian Mounted Police' special working conveying personnel and horses from Edinburgh Waverley to Dundee on 15th June 1957.

To the south of Dalmeny, near Kirkliston, preserved veteran Caledonian Railway 4-2-2 No. 123 of 1886 passes Queensferry Junction on the Edinburgh Waverley to Glasgow Queen Street main line en route to Dunfermline with the 'Lincolnshire Poacher', an enthusiasts' special train on 12th June 1960. Built for exhibition at Edinburgh, No. 123 was unique in being the sole engine of the class and the only 4-2-2 type to run in Scotland. During the 'Race to the North' in 1888, the engine worked express passenger trains from Carlisle to Edinburgh Princes Street in under two hours.

West Lothian rail tours

At the same location on 12th June 1960, a joint Railway Correspondence & Travel Society and Stephenson Locomotive Society 'Scottish Railtour' special train heads north behind the restored North British Railway 'Glen' 4-4-0 No. 256 *Glen Douglas*.

At Bo'ness station, the 1.57 pm branch passenger train for Polmont waits to leave the terminus behind Class C16 4-4-2T No. 67488. Built in 1916 as North British Railway No. 444 this engine was withdrawn from Polmont shed in October 1959. The four-mile long branch was closed to passenger traffic shortly after this picture was taken on 31st March 1956. The expanding area to the east of the former North British Railway station is being redeveloped by the Scottish Railway Preservation Society which was formed in November 1961 to save, restore and operate historical items representative of the Scottish railway scene.

Bo'ness to Polmont branch

En route to Polmont on the same working ten days later, Class J37 0-6-0 No. 64636 passes Kinneil Colliery, between Bo'ness and Manuel, the branch junction point on the Edinburgh Waverley to Glasgow Queen Street main line between Linlithgow and Falkirk High.

Privately owned by the Edinburgh Corporation Gas Department, green liveried Andrew Barclay, 0-4-0ST No. 7, builders No. 1036 built at Kilmarnock in 1904, was photographed between shunting duties at Granton depot, Edinburgh on 18th May 1956.

Barclay industrial locomotives in the Lothians

Andrew Barclay 0-4-0ST, No. 917 built in 1902, poses for the camera at Bo'ness circa 1957. Transferred to Fisons Ltd in 1948, the engine previously worked at Locharbriggs Quarry, Dumfries and was scrapped in 1961.

Edinburgh Gas Department 0-4-0ST No. 10, Andrew Barclay No. 1890 of 1926, was also recorded at Granton, Edinburgh in May 1956. Rebuilt at Granton in 1952, the green liveried engine spent its last days in private company ownership and is now preserved.

Footplate staff pause during their cleaning of Class C16 4-4-2T No. 67492 at North Berwick shed. The engine was being prepared for a Stephenson Locomotive Society special tour train on 26th August 1958. Built by the North British Locomotive Company in 1916 as NBR No. 448, it was sub-shedded at North Berwick in the 1950s and was withdrawn from Edinburgh St Margaret's depot in March 1960.

North Berwick

No. 67492 again features in this general view of the engine shed at North Berwick. A Class C16 4-4-2T and a Class V3 2-6-2T were officially allocated to North Berwick in 1954. The 4½-mile long branch line from Drem, on the East Coast Main Line, although now no more than a long siding, still retains its passenger services to Edinburgh Waverley.

Gullane branch

The 6½-mile long former North British Railway branch line to the holiday and golfing resort of Gullane, situated on the Firth of Forth, left the East Coast Main Line at Aberlady Junction, a mile west of Longniddry. On 21st June 1958, a Sunday school special rail trip ran to the coastal terminus which was closed to passenger traffic in September 1932. Freight traffic continued on the branch until being withdrawn in 1964. The excursion train was photographed on arrival at Gullane station, headed by St Margaret's depot-based Class V3 2-6-2T No. 67624.

Haddington branch

On 11th June 1960, a special train organised by the Stephenson Locomotive Society travelled over former passenger lines in East Lothian. On the climb towards Haddington, Class J35/4 0-6-0 No. 64489 heads the four-coach train away from the East Coast Main Line junction at Longniddry. The former North British Railway branch line was closed to passenger traffic in 1949 and was closed completely on 1st April 1968.

Only 17 miles from her destination at Edinburgh Waverley, Class A3 4-6-2 No. 60057 *Ormonde* speeds the down 'Flying Scotsman' through Longniddry on the East Coast Main Line on 24th May 1957. Named after the 1886 Derby and St Leger winner, Edinburgh Haymarket shed's No. 60057, a Doncaster-built Pacific, LNER No. 2556, was withdrawn from service in October 1963.

Longniddry

Headed by Class D11/2 4-4-0 No. 62690 *The Lady of the Lake*, a line of four 'Directors', Nos 62677 *Edie Ochiltree*, 62693 *Roderick Dhu* and 62694 *James Fitzjames*, await their ultimate fate in a siding alongside the East Coast Main Line at Longniddry on 12th June 1958.

Berwick, Roxburgh and Selkirk

The former North British Railway line between Reston, on the East Coast Main Line north of Berwick-upon-Tweed, and St Boswells, ran west across Berwickshire via Chirnside, Duns, Greenlaw and Earlston before reaching the Edinburgh to Carlisle Waverley route at Ravenswood Junction. The line crossed the River Tweed on the 19-arch Leaderfoot Viaduct which has recently undergone restoration and repair work. The signalman at Chirnside receives the tablet from the crew of Class J39 0-6-0 No. 64843 as it slows through the station with the Duns freight train on 23rd April 1960.

Chirnside and Duns

As a result of severe flood damage in the Berwickshire area, a section of the line between Duns and Greenlaw was closed to passenger traffic in August 1948. Closure of the Duns to Reston section followed in 1951, its goods service however, survived until 1966. The photograph shows a general view of Duns station looking west on 23rd April 1960.

The 10½-mile long Lauder Light Railway was opened in July 1901 and ran from Fountainhall, on the Waverley route, to Lauder via Oxton. With its severe weight restrictions and a speed limit of 25 mph, the branch line duties were shared in British Railways days by two ex-Great Eastern Railway Class J69 0-6-0Ts which, to spread the load, ran with empty side tanks, their water supply being carried in tenders coupled to the engines.

The line lost its passenger services in September 1932 and was closed to goods traffic in October 1958. The photograph shows the scene at Lauder station on a dull 1st November 1958 as British Railways Standard Class 2MT 2-6-0 No. 78049 is surrounded by enthusiasts and local dignitaries after working a Branch Line Society special train.

Berwickshire branch lines

A Branch Line Society rail tour covering Border area lines was run on 4th April 1959 behind Class D34 4-4-0 No. 62471 *Glen Falloch*. The special is seen at Gordon station on the single line between Greenlaw and Earlston before continuing its journey to St Boswells, Roxburghshire on the Waverley route, between Carlisle and Edinburgh.

Following the south bank of the River Tweed, the cross-country line from St Boswells, on the Waverley route to Tweedmouth on the East Coast Main Line, served stations at Maxton, Rutherford, Roxburgh Junction and Kelso. East of Kelso the line continued in former North British Railway territory serving stations at Sprouston and Carham before reaching the Northumberland border where the North Eastern Region of British Railways took over the services. The Branch Line Society tour train is seen again at St Boswells station, Roxburghshire, after arriving from Gordon and Earlston. No. 62471 *Glen Falloch* is about to leave on the next stage of its Border journey to Roxburgh Junction.

Roxburghshire rail tour

After arrival at Roxburgh Junction, No. 62471 *Glen Falloch*, looking smart in its British Railways mixed traffic black livery lined out in red, cream and grey, stands alongside a vintage North British Railway lower quadrant signal before leaving for Jedburgh with the Branch Line Society rail tour on 4th April 1959.

The 7-mile long branch line to Jedburgh, which followed the River Teviot southwards, ran from Roxburgh Junction on the cross-country line between St Boswells and Berwick-upon-Tweed, serving stations at Kirkbank, Nisbet and Jedfoot. Class J37 0-6-0 No. 64624 piloting the preserved North British Railway 'Glen' 4-4-0 No. 256 *Glen Douglas*, were recorded at Jedburgh terminus with a special tour train on 9th July 1961. The Jedburgh branch was closed to passengers in August 1948 and to all traffic in August 1964.

Jedburgh branch

Heading a Branch Line Society and Stephenson Locomotive Society 'Easter Rambler' rail tour train, Class B1 4-6-0 No. 61324 was photographed at Jedfoot with the special working on 14th April 1963.

The steeply graded and scenic Waverley route which served the former North British Railway territory between Edinburgh and Carlisle, ran through Midlothian and the Border country shires of Selkirk and Roxburgh before crossing the border into England at Kershope Burn. Class A1 4-6-2 No. 60118 *Archibald Sturrock* is seen at the isolated outpost of Riccarton Junction, only accessible by rail, between Hawick and the border, with a freight train from Edinburgh Millerhill to Carlisle on 14th August 1963.

Waiting to leave the junction with a freight working to Edinburgh is Class V2 2-6-2 No. 60957, the call being made to allow the locomotive crews to change over between the two trains. Not without much local opposition, the Waverley route was closed completely on 6th January 1969.

The Waverley Route

Snow was often an operational hazard on the more exposed sections of the line. Class J38 0-6-0 No. 65914 was recorded in charge of a snow clearance train near Whitrope Summit, 1,006 feet above sea level, between Newcastleton and Shankend during a spell of severe winter weather in January 1963 when the line was totally blocked.

The locomotive depot at Hawick, situated on a slightly lower level alongside the station, had an allocation of 22 steam engines in 1954, nearly all of which were of North British Railway origin. The exception was a Class Y1 0-4-0T, No. 68138 of 1927 LNER vintage, which was sub-shedded at Kelso. Class C16 4-4-2T No. 67489 is seen at the depot in company with other ex-North British Railway locomotive types on 19th September 1955.

Hawick shed

The driving wheel splasher of Hawick shed's Class D30/2 'Scott' 4-4-0 No. 62432 *Quentin Durward* showing the hand-painted name. All engines of this Reid-designed class carried names associated with characters which appeared in Sir Walter Scott's novels.

Border winter

After a bitterly cold winter's day employed on snow clearance duties in the area on 8th February 1958, Class J36 0-6-0 No. 65327 of Edinburgh St Margaret's depot, (sub-shedded at Galashiels), poses for the camera at Galashiels, Selkirkshire. Built at Cowlairs Works in 1900, this former North British Railway engine remained in service until withdrawal from Thornton Junction shed in November 1965.

Dumfries, Kirkcudbright and Wigtown

Class 2P 4-4-0 No. 40614 stands in the bay platform at the west end of Dumfries station with an afternoon local passenger working to Stranraer on 13th June 1959. Alongside the local train is the preserved Great North of Scotland Railway 4-4-0 No. 49 *Gordon Highlander* which had arrived earlier in the day with a rail tour special working from Glasgow.

Dumfries

A general view of Dumfries locomotive depot photographed from the Annan road overbridge on 14th April 1957. In 1954, the shed had an allocation of 38 steam engines. Seen from left to right are: a British Railways Standard Class 4MT 2-6-0, a Stanier Class 5MT 4-6-0, an ex-Caledonian Railway 0-6-0, a Stanier Class 8F 2-8-0, a visitor from Carlisle Kingmoor depot and an ex-Caledonian Railway 0-6-0T.

At Dumfries station, the preserved Great North of Scotland Railway 4-4-0, No. 49 *Gordon Highlander*, marshals coaching stock including observation car No. SC281, in preparation for the return journey to Glasgow via Kilmarnock with a Stephenson Locomotive Society 'Golden Jubilee' special rail tour on 13th June 1959. On its first main line run since restoration, the 4-4-0 worked the train from Glasgow St Enoch Station via Carstairs Junction on the West Coast route and over the disused line between Lockerbie and Dumfries.

Displaying her Carlisle (Kingmoor) depot on the bufferbeam, Class 4F 0-6-0 No. 44189, deputising for the regular Stanier Class 3MT 2-6-2T, waits to leave Dumfries with the morning Kirkcudbright branch passenger train in the summer of 1954. The final passenger service on the branch ran on 1st May 1965.

The West Coast Main Line over Beattock

The gruelling 10-mile long 1 in 75 climb from Beattock station to Beattock Summit, 1,014 feet above sea level, in the Lowther Hills mid-way between Carlisle and Glasgow, was a severe test for locomotives and their crews in the days of steam. Day and night, in fair weather and foul, the Anglo-Scottish trains did battle with the gradient on the West Coast Main Line, and provided the railway enthusiast with the spectacular sight and sound of steam engines hard at work. Resplendent in clean British Railways maroon livery, Stanier Class 8P 'Duchess' 4-6-2 No. 46225 *Duchess of Gloucester* crosses the shire border between Dumfriesshire and Lanarkshire with the 11.15 am Birmingham New Street to Glasgow Central express on 9th July 1960. Withdrawn from service in September 1964, No. 46225 was later cut up for scrap at the West of Scotland Shipbuilding Company, Troon, Ayrshire.

On the descent of Beattock Bank, Class 8P 'Princess Royal' 4-6-2 No. 46201 *Princess Elizabeth* speeds down the incline to the south of Greskine signal box with the 1.30 pm Glasgow Central to London Euston express – the 'Mid-Day Scot' – on 14th August 1958. Famous for its record-breaking exploits in November 1936 on 'non-stop' test runs from Euston to Glasgow and back, *Princess Elizabeth* was saved from the scrap heap in 1961 and has been restored in working order to its original LMSR condition by the 6201 Princess Elizabeth Locomotive Society. In July 1958, 'Princess Royal' class Nos 46201 and 46210 were transferred to the Scottish Region from the London Midland Region in exchange for 'Duchess' class Nos 46220 and 46221.

Hauling a passenger train composed of ex-LMSR coaching stock including a twelve-wheel dining car, British Railways Standard Class 6 'Clan' 4-6-2 No. 72007 *Clan Mackintosh* tackles the tree-lined climb of Beattock Bank near Harthope with the morning Manchester Victoria and Liverpool to Glasgow Central express on 7th July 1958.

One of Beattock shed's banking engines, Class 4MT 2-6-4T No. 42130, provides uncoupled rear end assistance to a northbound freight train near Auchencastle on the 10-mile long drag to Beattock Summit on a spring morning in 1958.

Maximum effort on the ascent of Beattock Bank, from Class 5MT 2-6-0 No. 42803 of Grangemouth depot, as she forges up the gradient and darkens the sky near Harthope on the border between Lanarkshire and Dumfriesshire with a Carlisle to Grangemouth freight train on 6th July 1958. The 'Crab' was broken up for scrap at the Motherwell Machinery & Scrap Company at Wishaw in December 1966.

Unassisted in the rear by a banking engine, a work-stained Crewe North-based Class 8P 'Princess Royal' 4-6-2, No. 46209 *Princess Beatrice*, toils uphill towards Beattock Summit near Harthope Viaduct with a 'relief' Birmingham New Street to Glasgow Central express on 9th July 1960.

Descending the incline at speed, Class 6P 'Jubilee' 4-6-0 No. 45711 *Courageous* of Glasgow Corkerhill depot is 'panned' by the camera between Greskine and Auchencastle on the lower reaches of Beattock Bank whilst heading a Glasgow Central to Liverpool and Manchester express in April 1958.

The 2-mile long Beattock to Moffat branch line was occasionally used for the quiet and secure overnight stabling of the Royal Train. After a Royal visit to the Scottish Central Lowlands on 7th July 1958, Class 5MT 4-6-0 No. 44902 was attached to the rear of the southbound train at Beattock station and the ensemble is seen leaving the West Coast Main Line and heading towards Moffat.

Royal train at Beattock

As daylight fades, the Royal Train slowly negotiates the sharp curve to the north of Beattock station and proceeds along the branch line past a small group of onlookers. Security was rather more relaxed in the 1950s!

Former Caledonian Railway Class 2P 0-4-4T No. 55164 was recorded by the camera on 10th July 1958 at Moffat station terminus after arriving from Beattock to pick up local freight. In early British Railways days, the short single-line branch was worked by an ex-London & North Western Railway Whale designed steam rail motor of 1905, which was later replaced by an ex-LNWR Webb push-pull fitted 2-4-2T, No. 46656. The branch was closed to passenger traffic in December 1954 and prior to complete closure in April 1964, the Class 4MT Fairburn 2-6-4T engines allocated to Beattock shed handled the remaining services.

Moffat branch

Six ex-Caledonian Railway Class 2P 0-4-4Ts were based at Beattock shed in 1954 for banking northbound trains to Beattock Summit and for use on the Moffat branch line workings. In the summer of 1954, No. 55187, painted in BR lined black livery and lettered 'British Railways' on its tank sides, was photographed outside the shed. The 0-4-4T was broken up for scrap at Kilmarnock Works in May 1955.

At Lochanhead, Kirkcudbright, on the Dumfries to Stranraer main line between Southwick and Dalbeattie, Class 6P 'Jubilee' 4-6-0 No. 45588 *Kashmir*, specially cleaned for the occasion, heads westwards with the 'Scottish Rambler' Easter rail tour on 15th April 1963.

South-West Scotland special trains

The former Glasgow & South Western Railway line from Newton Stewart to Whithorn witnessed steam activity once again when a joint Stephenson Locomotive Society and a Branch Line Society 'Scottish Rambler' rail tour visited the rural Wigtownshire branch line on 15th April 1963. The train of open wagons behind Class 2F 0-6-0 No. 57375, the last survivor of the class, is seen at Garlieston, a tidal port on Wigtown Bay, at the end of a short spur line from Millisle, between Wigtown and Whithorn. Until the 1930s, Garlieston was used by boat trains which connected with day-trip steamer services to the Isle of Man – the shortest sea crossing from Scotland.

One of the last surviving Scottish-based Class 4P 3-cylinder Compound 4-4-0s to remain in service, No. 40920 of Ayr depot, leaves Stranraer Town station with an early morning passenger train to Ayr and Glasgow St Enoch on 3rd August 1956.

Stranraer

A general view of Stranraer Harbour station on 14th June 1963. Waiting with the connecting passenger train service from the Larne steamer *Caledonian Princess*, Class 5MT 4-6-0 No. 44723 is about to leave the terminus with the Irish boat train for Glasgow St Enoch.

Lanark, Dunbarton, Renfrew and Ayr

Central

Glasgow Central station, the former 17-platform Caledonian Railway terminus covering 13 acres in the heart of the city on the north bank of the River Clyde, is the busiest station in Scotland. Until the early 1960s, steam power dominated on the West Coast Main Line on Anglo-Scottish expresses from Glasgow, local commuter and Cathcart Circle underground trains, together with intensive rail services to Paisley and the Clyde coast towns. Additional traffic was gained on the closure of St Enoch station in June 1966. In British Railways maroon livery, Class 8P 'Duchess' 4-6-2 No. 46248 *City of Leeds* is about to leave Central Station with the 5.45 pm express to London Euston on 17th August 1958.

Glasgow stations

Buchanan Street

Buchanan Street station replaced the old Glasgow, Garnkirk & Coatbridge station at Townhead. The ex-Caledonian Railway five-platform terminus provided rail services to Aberdeen, Dundee, Stirling, Perth, Oban and Inverness. The last regular steam-hauled passenger services to run in Scotland, the three-hour expresses to Aberdeen, were hauled by the surviving Class A4 4-6-2s. No. 60019 *Bittern* is seen about to leave Buchanan Street station with the last 'official' Class A4 trip to Aberdeen on 3rd September 1966. The regional Scotrail HQ now stands on the site of the terminus which was closed in that year and the building is seen under construction to the right of the picture.

Queen Street High Level

Glasgow Queen Street High Level station was the 1842 terminus of the original 46-mile long Edinburgh & Glasgow Railway. As well as linking Scotland's major cities, the oldest of Glasgow's main stations served Edinburgh and the East Coast Main Line, Dundee, Aberdeen, Inverness, the West Highland lines to Fort William, Mallaig, Oban and numerous local destinations. On 24th April 1961, Class V1 2-6-2T No. 67664 is about to start its journey from Platform 8, through the tunnel and up the steep 1 in 41 incline towards Cowlairs with a stopping passenger train to Kirkintilloch.

Queen Street Low Level

Glasgow Queen Street Low Level underground station served the city's western suburbs by rail north of the River Clyde and eastwards to Airdrie. Class V1 2-6-2T No. 67601 emerges briefly from the gloom into the daylight below Queen Street West signal box which still retained its original North British Railway signals when this picture was taken on 20th April 1957. The last steam-hauled passenger train to use the station ran to Helensburgh on 4th November 1960.

Polmadie

With an allocation of 178 steam engines in April 1954, Glasgow Polmadie depot, situated on the former Caledonian Railway main line between Glasgow Central and Rutherglen stations, always had a wide variety of express passenger and freight motive power on show. One of the first Class 8P 'Duchess' 4-6-2 Pacific engines to be withdrawn in early 1963 was No. 46246 *City of Manchester*. Still retaining its sloping smokebox, a legacy of its streamlined days, the Stanier-designed engine is seen at Polmadie taking water surrounded by ex-Caledonian Railway freight engine types before working the southbound 'Mid-Day Scot' on 22nd April 1956.

Glasgow shed scenes

Corkerhill

At Corkerhill depot, situated between Bellahouston and Mosspark stations to the south west of the city, British Railways Standard Class 6 'Clan' 4-6-2 No. 72005 *Clan Macgregor* has just arrived 'on shed' for servicing after working a parcels train over the former Glasgow & South Western Railway line from Carlisle via Dumfries and Kilmarnock in July 1959.

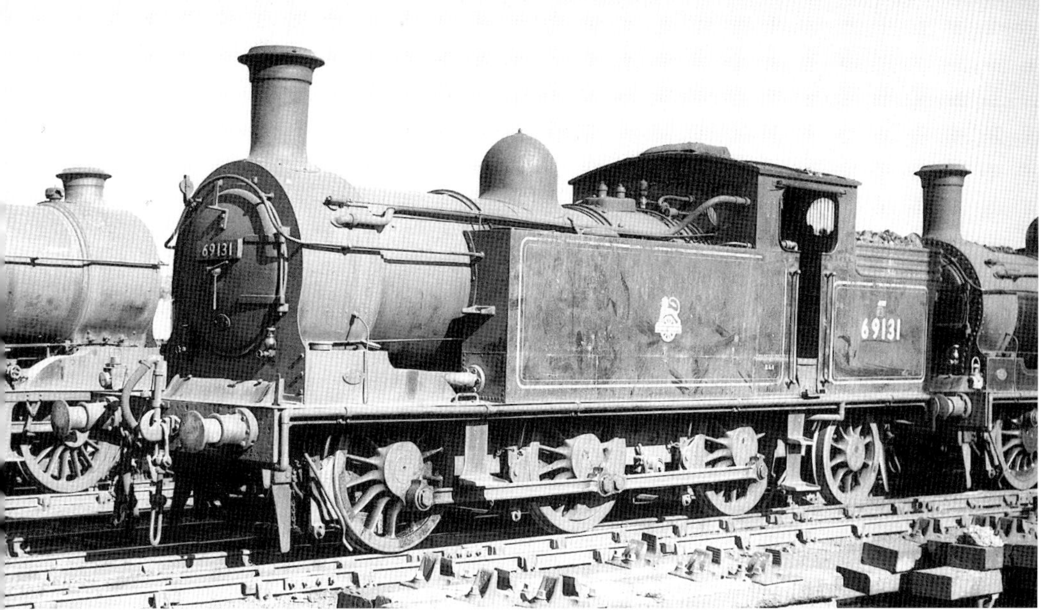

Eastfield

Situated at the top of Cowlairs incline, 1½ miles from Glasgow Queen Street station, the 14-road Eastfield depot built by the North British Railway in 1904, had the city's second largest allocation of 139 steam engines in 1954. Spending most of its life providing rear end assistance to east-bound passenger trains on the climb out of Queen Street station terminus, Class N15/2 0-6-0T No. 69131, one of the Cowlairs incline banking engines of 1910 and fitted with slip-coupling apparatus, was recorded in clean condition at Eastfield depot on 10th June 1956. The engine was later sold to the Cambus Atlas Steel Works, Clarkston for scrap.

Kipps (Coatbridge)

Two Class J36 0-6-0s, Nos 65285 and 65287 based at Kipps depot near Coatbridge Sunnyside station were fitted with cut-down boiler mountings for working under a low bridge on the former Caledonian Railway main line at Garnqueen North Junction to gain access to the siding serving Gartverrie brick works. No. 65285 with short chimney and dome, was photographed at Kipps shed on 13th May 1956.

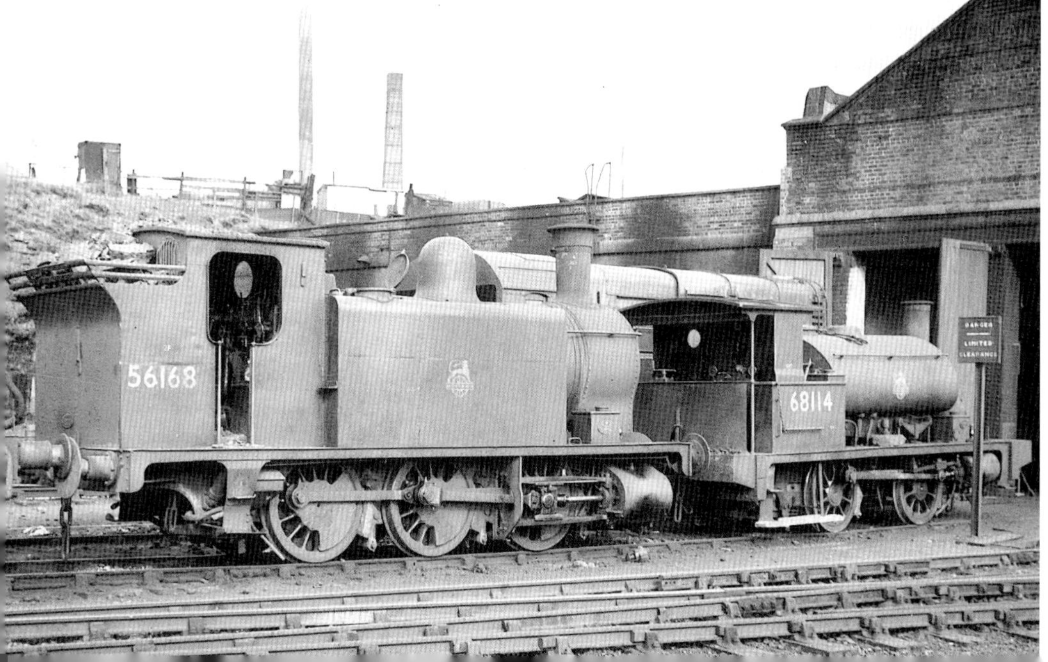

Dawsholm

To the north east of Glasgow, and north of Maryhill station, the former Caledonian Railway engine shed at Dawsholm had an allocation of 50 steam engines in 1954. On 19th August 1958, Class 2F 0-6-0T No. 56168 and Class Y9 0-4-0ST No. 68114 repose at the depot. Prior to its closure in 1964, Dawsholm shed was the storage and maintenance point for the four preserved Scottish pre-Grouping locomotives – Caledonian Railway 4-2-2 No. 123, Highland Railway 4-6-0 No. 103, North British Railway 4-4-0 No. 256 *Glen Douglas*, and Great North of Scotland Railway 4-4-0 No. 49 *Gordon Highlander*.

At Peacock Cross on the former North British Railway line between Shettleston and Hamilton, Lanarkshire, Class N15/1 0-6-2T No. 69178 was engaged on shunting activities amidst typical industrial surroundings when photographed in April 1962.

North British Railway tank engines around Glasgow

North British Railway Class J88 0-6-0T No. 68345 fitted with a tall stovepipe chimney is seen at Glasgow Eastfield depot on 10th June 1956 in company with ex-North Eastern Railway Class J72 0-6-0T No. 68709 which has received the same treatment.

Class 4MT 2-6-4T No. 42173 waits for custom at Lanark station terminus with a two-coach local passenger train to Carstairs Junction in April 1960. Lanark also provided rail services to Muirkirk, Ayrshire, and Glasgow Central station.

Lanark and Carstairs Junction

A specially turned out Class 4MT 2-6-4T, No. 42059 of Glasgow Polmadie depot, heads west through Carstairs Junction with an officers' saloon coach special working on 17th September 1956. Originally West Coast Joint Stock, the saloon No. SC45018M, incorporating LNWR features, was rebuilt with an observation end to allow engineers to inspect the installation of overhead lines for Glasgow's suburban electrification. On withdrawal, the saloon was restored to Caledonian Railway livery as No. 41 and has been used recently on the luxury 'Royal Scotsman' touring train.

One of Glasgow Polmadie's 1954 allocation of nine Class 8P 'Duchess' 4-6-2s, No. 46232 *Duchess of Montrose*, coasts through Carstairs Junction with the down 'Royal Scot' on the final leg of the train's seven-hour journey from London Euston to Glasgow Central on 16th August 1958.

On the same day, the 11.5 am Birmingham New Street to Glasgow Central express pulls strongly away from its Carstairs Junction stop with Class 8P 'Princess Royal' 4-6-2 No. 46201 *Princess Elizabeth* in charge; a scene now dominated by 25kV overhead catenary lines.

Carstairs Junction shed's Class 2P 0-4-4T No. 55261, one of a post-Grouping development of ten engines built for the LMSR by Nasmyth, Wilson in Manchester, in 1925, has been commandeered from its regular station 'pilot' duties to work an afternoon local passenger train to Lanark and makes a smoky departure from Carstairs station on 17th September 1956.

With Carstairs Junction locomotive depot in the background and the Edinburgh Princes Street connecting line curving eastwards to the right of the picture, a Sunday track-laying train hauled by Class 3P 4-4-0 No. 54461 leaves the junction at Strawfrank and heads up the West Coast Main Line towards Symington in May 1957.

On a stormy day in July 1959, a brief glimpse of the sun illuminates Class 8P 'Duchess' 4-6-2 No. 46253 *City of St Albans*, displaying her tartan train name headboard, as she pounds up the last mile of Beattock Bank, unassisted in the rear, with the Glasgow Central bound 'Royal Scot'.

The Battle of Beattock

At the same location, and with the sound of her exhaust echoing around the Lowther Hills, ex-War Department 'Austerity' 2-10-0 No. 90762 struggles to lift a northbound freight on the final assault of the gradient to Beattock Summit on a still summer's evening in July 1959.

With fully open regulator, Class 6P 'Jubilee' 4-6-0 No. 45702 *Colossus* of Manchester Newton Heath depot, darkens the sky over Elvanfoot as she noisily attacks the rising gradient on the northern approaches to Beattock Summit with the 4.30 pm Glasgow Central to Manchester express on 10th April 1960.

A photograph taken from the guard's compartment window in the summer of 1956 as Class 4MT 2-6-4T No. 42192 drops away from the rear of a northbound express at Beattock Summit, 1,014 feet above sea level, after providing uncoupled banking assistance up the 10-mile long climb from Beattock station.

Class 6P 'Jubilee' 4-6-0 No. 45716 *Swiftsure* of Carlisle Kingmoor depot, coupled to a 3,500 gallon capacity Fowler tender, catches the late afternoon sun as she trundles past Elvanfoot on the northern approaches to Beattock Summit with a southbound freight on 16th July 1958.

Through the Clyde Valley

In the previous year, a summer holiday 'relief' passenger train from Glasgow Central to Blackpool accelerates along the West Coast Main Line after a signal check to the south of Crawford, behind unrebuilt Class 6P 'Patriot' 4-6-0 No. 45517 of Bank Hall depot, Liverpool, in clean British Railways Brunswick green livery.

Class 7P 'Royal Scot' 4-6-0 No. 46102 *Black Watch* was one of five engines of the class allocated to Glasgow Polmadie depot in 1954. It was in charge of the morning Glasgow Central to Liverpool and Manchester express when photographed at speed near Symington on the West Coast Main Line between Carstairs Junction and Beattock on 14th June 1958.

As shadows lengthen over the village of Elvanfoot, Lanarkshire, on the northern approaches to Beattock Summit, Polmadie's Class 8P 'Duchess' 4-6-2 No. 46222 *Queen Mary* heads into the night with the southbound 'West Coast Postal' on 9th July 1959.

A general view of the locomotive depot at Helensburgh, Dunbartonshire on 8th March 1958 with Class V3 2-6-2T No. 67619 'on shed'. A sub-depot of Glasgow Parkhead, the shed had an allocation of around ten steam engines in the mid-1950s, mostly the Class V3 2-6-2Ts which worked the Glasgow and Airdrie passenger services.

Helensburgh

Helensburgh Central Station photographed on the same day. Class V3 2-6-2T No. 67619 is about to depart with an eastbound passenger train, prior to the introduction of the Glasgow area 'Blue Train' electric services in the early 1960s.

Beside Loch Lomond

An excursion train from Glasgow, hauled by Class B1 4-6-0 No. 61333, has just arrived at Balloch Pier station terminus at the southern end of Loch Lomond in the summer of 1956. Under the watchful eye of the captain of the paddle steamer *Maid of the Loch*, the passengers are about to embark for a day's cruise to Tarbert and Ardlui, at the head of the loch. The small engine shed at nearby Balloch, a sub-depot of Glasgow Eastfield, had a 1954 allocation of one Class J36 0-6-0 and two Class V1 2-6-2T locomotives.

A front view of Class C15 4-4-2T No. 67460 showing the push-and-pull fittings. The local passenger train is about to leave Craigendoran station for Arrochar in the summer of 1957.

Arrochar push-and-pull

Part of the West Highland line from Glasgow Queen Street to Fort William, (the former North British Railway line from Craigendoran, on the Firth of Clyde), to Arrochar via Helensburgh Upper station was a 'push-and-pull' steam service operated by Class C15 4-4-2Ts Nos 67460 and 67474 of Eastfield depot, Glasgow. The last two 'Yorkie' survivors of the class are illustrated at work on the steeply graded line before their replacement by a diesel rail-bus in 1960, a service which lasted until the intermediate stations lost their passenger facilities on 14th June 1964. Nos 67460 and 67474 were built by the Yorkshire Engine Company Sheffield in 1912 and 1913 respectively. Both engines were withdrawn from service in April 1960.

The locomotive crew of No. 67474 in the cab of their Class C15 4-4-2T at Arrochar station before the return departure of the passenger service to Craigendoran.

Class C15 4-4-2T No. 67474 waits to leave Arrochar station with the Craigendoran passenger train on 21st March 1959.

Carrying express passenger code headlamps, Class C15 4-4-2T No. 67460 propels its two-coach train northwards between Helensburgh Upper station and Rhu, en route to Arrochar in the summer of 1957.

The subject of extensive rebuilding by the Caledonian Railway in 1903, the attractive terminal station at Wemyss Bay, Renfrewshire, provided a connection from Glasgow for rail passengers using the steamer services between the popular Clyde coast holiday resorts. The photograph shows the clean and cheerful aspect of the station in the 1950s. The ornate crests removed from former Clyde steamers are displayed on each side of the wooden floored approach way leading to the main station concourse.

Wemyss Bay
and
Greenock

A line up of four former Caledonian Railway Class 3P 'Caley Bogie' 4-4-0s stored out of use and never to run again, at Greenock Princes Pier depot in April 1956. Nearest the camera is No. 54506 and alongside are Nos 54479, 54492 and 54498. Officially classified as a sub-shed of Greenock Ladyburn, the depot had an allocation of twelve steam engines in 1954 including No. 47169, one of the five Fowler 'Dock Tanks' based on the Scottish Region.

Class 2F 0-6-0T No. 47169 poses for the camera during shunting operations at Greenock Princes Pier station on 16th April 1956. A 1928 Fowler design for the LMSR, the 'Dock Tank' was known locally as 'Wee Maggie', and the painted name is just visible at the top of the smokebox door.

Class 0F 'Caley Pug' 0-4-0ST No. 56031, ex-Caledonian Railway No. 622, attached to a wooden tender, shunts a variety of stock in the bay platform at the east end of Greenock Central station in April 1956.

A brake van tour train run by Glasgow University Railway Society heads past Glenfield station on the ex-Caledonian Railway line between Paisley St James and Lyoncross, behind Class 2F 0-6-0 No. 57268 on 15th May 1961.

Paisley and Renfrew Wharf

The last train about to leave Renfrew Wharf for Glasgow Central station on 28th April 1967, the final day of working, behind British Railways Standard Class 4MT 2-6-4T No. 80004.

Heads of Ayr

Purchased by Butlins on withdrawal from British Railways' service as an added attraction at their holiday camp at Heads of Ayr, Class 8P 'Duchess' 4-6-2 No. 46233 *Duchess of Sutherland* contrasts with ex-London, Brighton & South Coast Railway 'Terrier' 0-6-0T No. 32662, formerly No. 62 *Martello*, built at Brighton in 1875. The engines, restored to their original liveries, were photographed on display at the camp in 1964.

'Duchess' at Barony Junction

Stanier Class 8P 'Duchess' 4-6-2 No. 46224 *Princess Alexandra* passes Barony Junction, Mauchline, on the former Glasgow & South Western Railway line between Old Cumnock and Kilmarnock on 11th June 1963. The train is the 11 am four-coach passenger working from Carlisle to Glasgow St Enoch via Dumfries.

En route to Ayr, the first engine of the British Railways Standard Class 6 'Clan' 4-6-2s, No. 72000 *Clan Buchanan* hurries a parcels train from Glasgow St Enoch through Barassie station on the former Glasgow & South Western Railway line between Irvine and Ayr in the summer of 1961. The engine was officially named on 16th January 1952 at Glasgow Central Station by the Lord Provost of Glasgow.

Steam at Barassie, Ayrshire

Class 5MT 2-6-0 No. 42737, running tender first, negotiates the point work on the approaches to Barassie station and takes the Kilmarnock line with a mixed freight working in 1961. Extensive wagon repair facilities existed at the wagon works nearby.

The former Glasgow & South Western Railway locomotive works at Kilmarnock, Ayrshire, was used for the light repair and cutting up for scrap of withdrawn Scottish Region locomotives. Its tender still lettered 'British Railways', Class 3P superheated 'Dunalastair IV' 4-4-0 No. 54445, formerly Caledonian Railway No. 117 built in 1912, awaits breaking up in the works yard in July 1952.

Kilmarnock Locomotive Works

Awaiting a similar fate at the works in July 1952 was one of the last Pickersgill 'Wemyss Bay Tanks', Class 4P 4-6-2T No. 55361, ex-Caledonian Railway No. 955, built in 1917. The engine had received the British Railways black mixed traffic livery, lined out in red, cream and grey, and spent its last days on Beattock incline banking duties.

A Cumming design of 1918 for the Highland Railway, Class 4F 4-6-0 'Clan Goods' No. 57956 still retaining its former LMS ownership initials on the tender, awaits the breaker's torch outside Kilmarnock Works in July 1952. Other ex-Highland Railway locomotive types which survived into the early British Railways era were a few of the Class 2P 'Loch' and 'Ben' 4-4-0s, two Drummond Class 1P 0-4-4Ts, and two Class 4P 'Clan' 4-6-0s, the last of which, No. 54767 *Clan Mackinnon* was scrapped at Kilmarnock in 1950. Apart from 'Jones Goods' 4-6-0 No. 103, no ex-Highland Railway engine survives. No. 54398 *Ben Alder* was stored pending possible preservation during the 1950s and early 1960s, but being in rebuilt form, it was not considered as part of the National Collection, and was dispatched to the scrap yards.

Running tender first, British Railways 'Standard' Class 4MT 2-6-0 No. 76108 pauses at Newton-on-Ayr station with the 8.5 am Ayr to Kilmarnock passenger train on 13th June 1963.

British Railways Standard 2-6-0s in Ayrshire

A mile-long branch line ran north east from Brackenhill Junction on the Dumfries to Kilmarnock line between Mauchline and Cumnock, to the village of Catrine, Ayrshire. British Railways Standard Class 2MT 2-6-0 No. 78026 heads the branch goods at Catrine on 6th April 1962. Freight services on this short branch line were terminated on 6th July 1964.

Ayrshire freight

Class B1 4-6-0 No. 61243 *Sir Harold Mitchell* of Ayr depot passes Dailly station south of Maybole with the Falkland Junction, Ayr to Girvan goods train on 22nd May 1963.

Stirling, Clackmannan, Fife, Kinross and Angus

A scene at the south end of Stirling station featuring one of the ubiquitous Stanier Class 5MT 4-6-0s, No. 44994, Horwich-built in 1947 and based at Edinburgh Dalry Road depot. It is about to leave with a stopping passenger train to Edinburgh Princes Street on a June day in 1961.

Stirling

Built by Neilson of Glasgow, former Caledonian Railway Class 2F 0-6-0 No. 57261, running tender first, trundles into the yards at the south end of Stirling station with a short local freight train in the summer of 1961.

On its 154-mile three-hour journey from Aberdeen to Glasgow Buchanan Street, Class A4 4-6-2 No. 60007 *Sir Nigel Gresley* accelerates away from its Stirling stop and passes the locomotive depot with the 1.30 pm express on 18th September 1964. This was the swansong of the Class A4 Pacifics and one of the last steam-worked passenger services in Scotland.

The 10.15 am passenger train to Edinburgh Princes Street about to leave Platform 3 at Stirling station behind Edinburgh, Dalry Road–based Class 5MT 4-6-0 No. 45183 in September 1964. Note the spick and span condition of the station.

Class 2P 0-4-4T No. 55126, one of the last survivors of the McIntosh Caledonian Railway 92 class of 1897, poses for the camera during station yard 'pilot' duties at Stirling on 1st September 1956.

Stirling station 'pilots'

The Stirling station 'pilot' Class 2P 0-4-4T No. 55195, an example of the McIntosh Caledonian Railway 439 class of 1900, shunts vans at the north end of the station on 1st July 1956.

Blue-liveried Caledonian Railway 4-2-2 No. 123 pauses at Throsk station, between Alloa and Larbert, with an Edinburgh & Lothians Model Railway Club special train on 2nd October 1959. Built for the 1886 Edinburgh International Exhibition, No. 123 was later used on directors' special workings and as a Royal Train 'pilot' engine, running ahead of the Royal Train to ensure a clear road.

Caledonian Railway No. 123 at work

An early appearance on the main line on 10th May 1958, after its restoration to working order, Caledonian Railway 4-2-2 No. 123 heads north out of Larbert with a trial trip hauling the two preserved Caledonian Railway passenger coaches.

The former North British Railway engine shed at Polmont was situated beside the Edinburgh Waverley to Glasgow Queen Street main line, between Linlithgow and Falkirk High stations. At the depot on 28th April 1957 were an ex-Great Eastern Railway Class J69/1 0-6-0T, No. 68524, the shed 'pilot' engine – a long way from its origins – and a Class 4MT 2-6-0, No. 43140 which is just leaving the shed to work a local, freight train. Polmont depot, with a 1954 allocation of 42 steam engines, was closed in May 1964.

Former North British Railway, Class J37 0-6-0 No. 64571, still displaying its early 'British Railways' lettering on the tender, stands at the west end of the shed on 21st April 1956.

Polmont shed

Built by the Yorkshire Engine Company, Sheffield in 1913, Class C15 4-4-2T No. 67472 returns to its home depot at Polmont after working a local passenger train from Grangemouth via Falkirk Grahamston on 31st March 1956.

Class C16 4-4-2T No. 67494 curves away from the main Edinburgh Waverley to Glasgow Queen Street line at Polmont with a two-coach passenger train for Falkirk Grahamston and Grangemouth on a summer's evening in 1958. Originally North British Railway No. 450, the 4-4-2T was withdrawn from service in 1961.

Polmont – Falkirk – Grangemouth local

Class C16 4-4-2T No. 67488 waits in the sidings adjoining Grangemouth station terminus before working a local passenger train back to Polmont via Falkirk Grahamston on 19th September 1955. The branch was closed to passenger operations in January 1965.

A Branch Line Society rail tour hauled by Class 3P 4-4-0 No. 54465, one of the last remaining Pickersgill designed 'Caley Bogies' of 1916, leaves Manuel Junction on the Edinburgh Waverley to Glasgow Queen Street main line between Linlithgow and Polmont, with the two restored brown and cream liveried Caledonian Railway coaches on 6th May 1960. The special train was about to head up the goods line to Avonbridge as part of a tour covering disused lines in this area.

Branch Line Society special train

The BLS special train, photographed after arrival at Manuel Junction from Bo'ness, behind a spotless No. 54465, formerly Caledonian Railway No. 121, displaying its 66B (Motherwell depot) shed plate. No. 54465 was earmarked as a possible candidate for preservation but the 4-4-0 was withdrawn from service in October 1962 and scrapped.

Class D11/2 'Director' 4-4-0 No. 62684 *Wizard of the Moor* pauses at Dollar station on the Devon Valley line, between Alloa and Kinross Junction, with a three-coach passenger train from Glasgow Queen Street to Perth General on 10th August 1957. The former North British Railway line was closed to passengers on 15th June 1964.

Dollar, Clackmannanshire

Lying in the shelter of the Ochil Hills, the derelict single-platform station at Dollar was photographed in September 1964, four months after its closure to passenger traffic.

Fife local

Class V3 2-6-2T No. 67675, based at Stirling Shore Road depot, makes a spirited exit from Rosyth Halt, Fife, with a local afternoon passenger train from Edinburgh Waverley to Stirling via Dunfermline and Alloa on 16th June 1958.

Lochty goods

Class J36 0-6-0 No. 65345, one of the last ex-North British Railway steam locomotives at work in Scotland, hauls a short goods train on the freight-only 14½-mile long line to Lochty, Fife, between Kennoway and Montrave, on 18th February 1961. The engine carries a small bufferbeam snowplough to remove mud from the rails at the many farm crossings on the line. Freight services on the Lochty branch were withdrawn in August 1964. The final section of the line, once owned by the Lochty Private Railway but now closed, was host to preserved Gresley Class A4 4-6-2 No. 60009 *Union of South Africa*. No. 65345 later became well-known for its appearance in the BBC television series 'Dr Finlay's Casebook'.

Only two Class D49/2 'Hunt' 4-4-0s, Nos 62743 and 62744, were allocated to Scottish Region in the 1950s. A very clean No. 62744 *The Holderness* of Dundee Tay Bridge shed, comes round the north curve past Inverkeithing Junction signal box and towards Inverkeithing station with a Dundee to Edinburgh Waverley stopping passenger train on 29th August 1956. No. 62744 was withdrawn from Hawick shed in December 1960.

Inverkeithing, Fife

A clean Class J37 0-6-0, No. 64636, hauls vintage coaching stock past a fine array of signals, on a Sunday school excursion train to the Fife coast, 20th June 1959. The train is returning to Glasgow Queen Street from Anstruther and hurries under the road bridge and through Inverkeithing station. Note the line-side notice warning of crossing the tracks when light is showing.

Clearly showing the original British Railways 'lion and wheel' emblem on the tender, Class K3/2 2-6-0 No. 61991 (Darlington-built in 1937 and based at Edinburgh St Margaret's depot), darkens the sky as she pounds slowly up the 1 in 70 gradient towards Hookhills cutting between Inverkeithing and North Queensferry on the approaches to the Forth Bridge with a southbound freight train on 26th May, 1956.

Climbing to the Forth Bridge

An express Aberdeen to London King's Cross meat train behind Dundee Tay Bridge-based Class V2 2-6-2 No. 60822, crosses Jamestown Viaduct north of Inverkeithing station on the northern approaches to the Forth Bridge on 25th April 1957.

Thornton Junction station, situated on the Edinburgh Waverley to Dundee Tay Bridge and Aberdeen main line between Kirkcaldy and Cupar, was also the busy junction point for westbound trains to Dunfermline, Fife Coast services to St Andrews via Leven, Elie, Anstruther, Crail and a freight line to Methil Docks. Thornton Junction depot, with a mid-1950s allocation of over 100 steam engines, mostly freight types, provided power for the heavy mineral traffic within the Fife coal fields.

The fireman of Class D11/2 'Director' 4-4-0 No. 62684 *Wizard of the Moor* of Glasgow Eastfield depot, attends to the coal supplies as his engine is prepared to work a Thornton Junction to Glasgow Queen Street stopping passenger train via the Forth Bridge and Dalmeny Junction in October 1956.

Thornton Junction, Fife

One of Dunfermline depot's three Class D30/2 'Scott' 4-4-0s, No. 62427 *Dumbiedykes*, heads round the sharp curve from Thornton Junction Station with the 5.5 pm local passenger train to Dunfermline Upper in July 1956.

On display at Dunfermline

Green-liveried former Cardowan Colliery, Fife 0-4-0ST No. 11, built by Gibb & Hogg of Airdrie in 1898, was presented to Dunfermline Carnegie Trust and is shown on display at Pittencrieff Park, Dunfermline in 1969.

Rumbling Bridge

At Rumbling Bridge, on the Devon Valley line between Dollar and Kinross, Class D34 'Glen' 4-4-0 No. 62490 *Glen Fintaig*, approaches a tall, North British Railway lattice-post signal with a three-coach passenger train from Perth to Glasgow Queen Street on 10th August 1957.

A rather grubby British Railways Standard Class 4MT 2-6-4T, No. 80123, arrives at Dundee Tay Bridge station with a morning local passenger train from Tayport, Fife, a journey which would have taken the train along the southern shores of the Firth of Tay and across the Tay Bridge.

Dundee

The ex-Caledonian Railway station terminus at Dundee West on 14th April 1956, as Class 4P 4-4-0 No. 40938 of Perth depot shunts empty coaching stock in the station yards. The station was closed to all traffic on 1st May 1965.

The elegant lines of Class C16 4-4-2T No. 67484 are seen to advantage in this photograph taken at Dundee Tay Bridge depot on 14th April 1956. The Dundee-based Class C16 4-4-2Ts were mostly employed on the former Dundee to Arbroath joint line passenger services terminating at Dundee East station, which was closed in early 1959, and on local passenger trains to Tayport, Leuchars Junction and St Andrews, Fife, via the Tay Bridge, and on station 'pilot' duties. No. 67484 was built by the North British Locomotive Company, Glasgow in 1915 and was withdrawn from Scottish Region service in 1960.

North British locomotives at Dundee

Class Y9 0-4-0ST No. 68100, one of six engines of the class allocated to Dundee during the mid-1950s, fitted with detachable spark arrester and coupled to a wooden tender, arrives for servicing at Dundee Tay Bridge depot after completing a spell of shunting duty at the docks on 14th April 1956.

Recently outshopped from Inverurie Works, Aberdeenshire, Class J37 0-6-0 No. 64574 of Dunfermline depot, Fife, is seen in the company of home-based Class C16 4-4-2T No. 67490 at the west end of Dundee Tay Bridge depot on 14th April 1956.

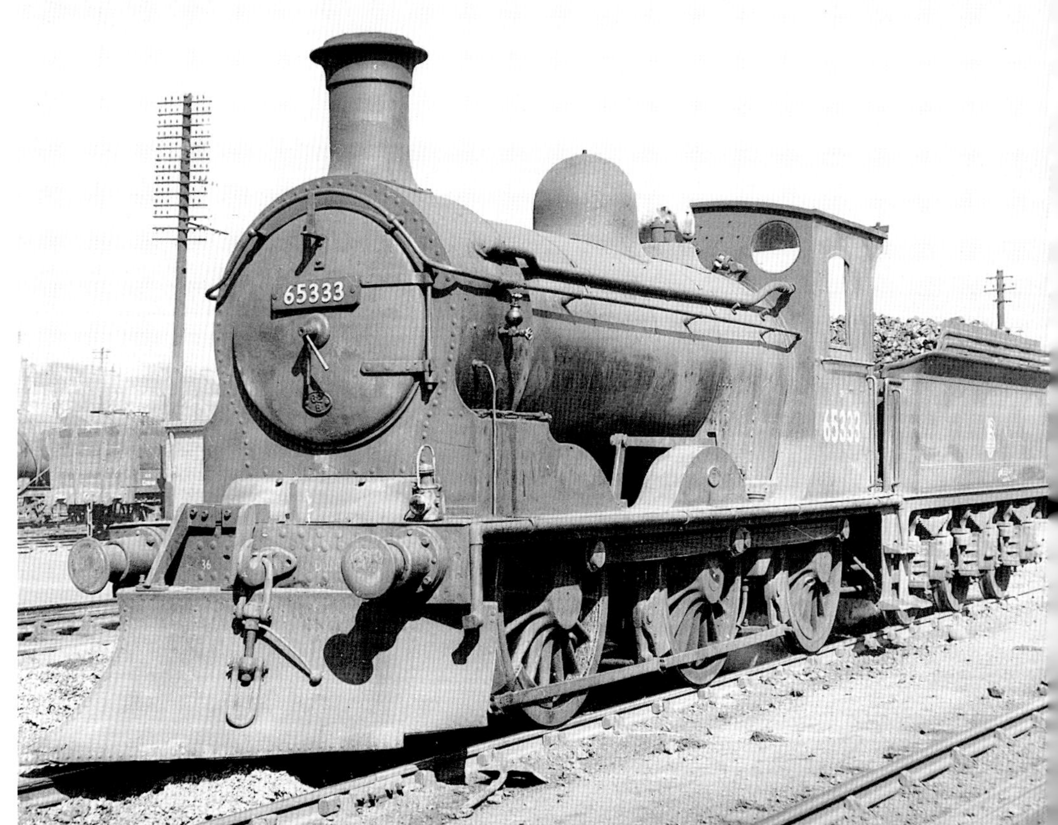

At its home depot on the same day, one of Dundee Tay Bridge depot's 1954 allocation of four Class J36 0-6-0s, No. 65333, (built in May 1900 at Cowlairs Works) fitted with a bufferbeam snowplough, simmers in the shed yards whilst awaiting its next call of duty.

Class 2F 0-6-0 No. 57441 was about to leave Kirriemuir station terminus, Angus, with a joint Stephenson Locomotive Society, Branch Line Society and Railway Correspondence & Travel Society 'Scottish Rail Tour' special train when photographed on 16th June 1960. The train includes the two restored Caledonian Railway passenger coaches next to the locomotive. The short Kirriemuir branch, which diverged northwards from the Strathmore main line west of Forfar, lost its passenger services in 1952 and was closed completely in 1965.

Branch lines in Angus

A wintry scene at Justinhaugh station on the former Caledonian Railway line between Forfar, Brechin and Bridge of Dun, as Class 3P 4-4-0 No. 54489 pauses with a pick-up goods train for Careston on 4th February 1961. The line remained open for freight traffic until 1967.

The Edzell branch goods with Class J37 0-6-0 No. 64587 in charge was photographed at Brechin station terminus on 4th March 1961. Situated on a short spur line between Bridge of Dun and Edzell, Brechin was closed to passenger traffic on 4th August 1952. Major refurbishment and restoration work is now in progress by the Brechin Locomotive Society.

Brechin and Edzell branches

On the same day, Class J37 0-6-0 No. 64587 halts at Stracathro with the Edzell branch daily goods from Brechin. The former Caledonian Railway branch line was closed to passenger traffic in September 1938.

Aberdeen, Kincardine, Banff and Moray

Exit from the Granite City

Fighting for adhesion on the wet rails, Class V2 2-6-2 No. 60955 passes the locomotive depot at Aberdeen Ferryhill on the climb out of Aberdeen with a southbound fitted freight on a gloomy summer's evening in June 1962.

Aberdeen's engine depots at Ferryhill and Kittybrewster had a combined allocation of 90 steam locomotives in April 1954. Twenty-two different class types were represented, ranging from the ex-Great North of Scotland Railway Class D40 and D41 4-4-0s to British Railways Standard Class 4 2-6-4Ts. As well as a variety of pre-Grouping examples, other interesting vintage locomotives were at work in the Aberdeen area, including an ex-Great Eastern Railway Class F4 2-4-2T, four ex-Great North of Scotland Railway Class Z4 and Z5 0-4-2Ts, an ex-North Eastern Railway Class G5 0-4-4T and six ex-North Eastern Railway Class J72 0-6-0Ts, one of which, No. 68700, was photographed in very clean condition at Kittybrewster depot in October 1956.

Locomotive contrasts at Aberdeen

A visitor from Perth depot, Class 6P 'Jubilee' 4-6-0 No. 45673 *Keppel*, in clean British Railways Brunswick green livery, (built at Crewe Works in 1935), stands on the turntable at Ferryhill shed in the summer of 1956.

The five-mile long branch line, opened in 1903 linking Fraserburgh and St Combs on the north eastern coast of Aberdeenshire, boasted a passenger service of twelve trains per day in each direction in 1963, serving halts at Kirkton Bridge, Philorth Bridge and Cairnbulg. Before dieselisation, the unfenced former Great North of Scotland Railway line was worked by a variety of steam engine classes including ex-Great Eastern Railway Class F4 2-4-2Ts fitted with cow-catchers, and Ivatt Class 2MT 2-6-0s, one of which, No. 46461, is about to leave the terminus at Fraserburgh for St Combs on 9th August 1956.

Aberdeenshire branch lines

Closed to passenger traffic in 1931, the disused branch from Inverurie, on the Aberdeen to Huntly line, to Old Meldrum was revisited on 21st April 1962 by a rail tour using the restored ex-Great North of Scotland Railway 4-4-0 No. 49 *Gordon Highlander*. The 4-4-0 was photographed at Old Meldrum station at the end of the 5¼-mile long branch.

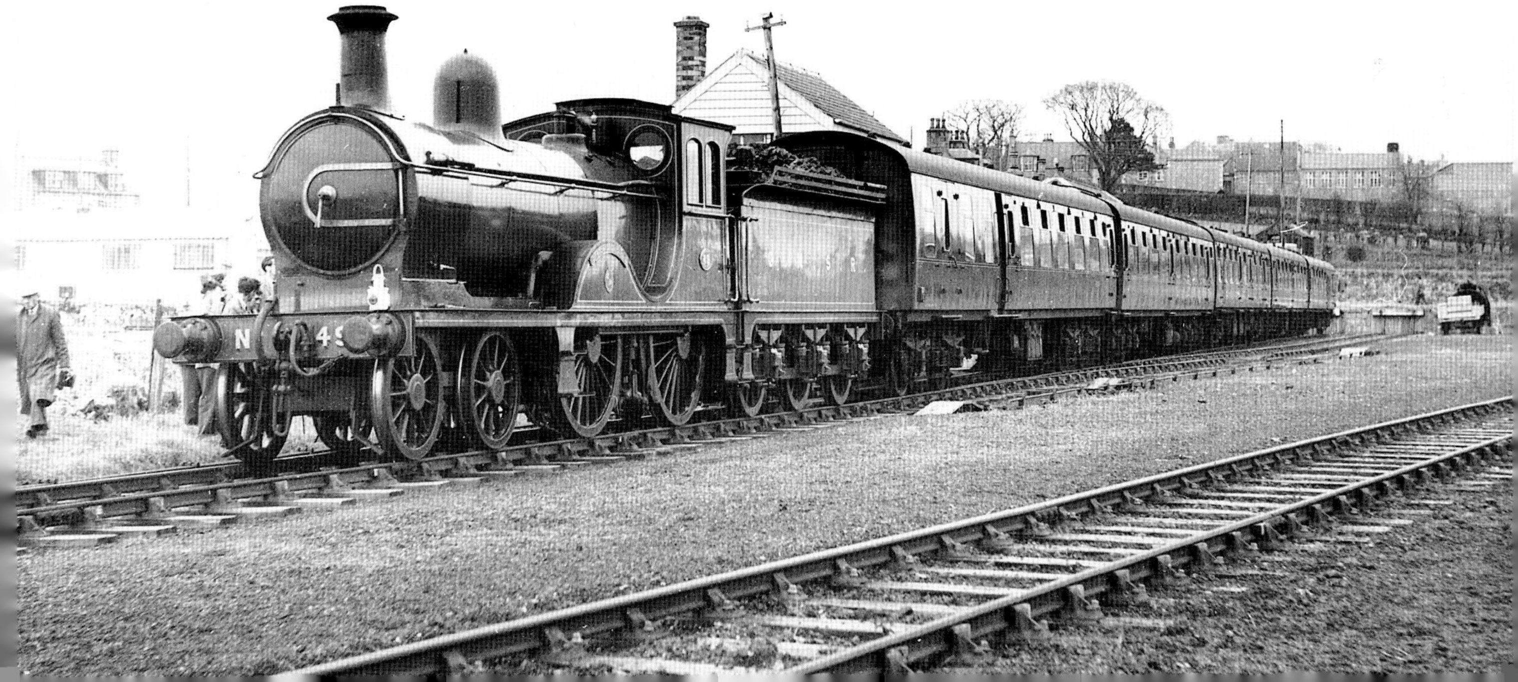

Closed to passenger traffic in January 1950 and finally abandoned on 3rd January 1966, the 16-mile long Alford to Kintore branch in Aberdeenshire was again steam worked on 13th June 1960 when the restored Great North of Scotland Railway 4-4-0 No. 49 *Gordon Highlander* was in charge of an enthusiasts' special train. The photograph shows the train in an attractive rural setting at the deserted Monymusk station, midway along the branch. In its heyday, the line also served stations at Kemnay, Tillyfourie, and Whitehouse.

Alford branch

No. 49 *Gordon Highlander* is seen again on the Alford to Kintore branch with the tour train at the disused Kemnay station on 13th June 1960.

Inverbervie branch

Heading north-east from Montrose, the coastal line to Inverbervie, Kincardineshire, saw a passenger train for the first time in many years when a week-long Scottish rail tour special train travelled over the former North British Railway branch line behind Class J37 0-6-0 No. 64615. The train is shown on 16th June 1960 at the intermediate station of Johnshaven, its platforms reverting to nature. The Inverbervie branch lost its passenger services in October 1951 and the line with its six stations closed completely in 1967.

Speyside local

Following the course of the River Spey, Class D40 4-4-0 No. 62271 storms out of Craigellachie. Banffshire, with the two-coach 2.55 pm passenger train to Boat of Garten via Grantown-on-Spey on 3rd October 1956. No. 62271, built at Inverurie Works in 1914, was withdrawn from service in November 1956, a month after this picture was taken.

One of the W. P. Reid designed ex-North British Railway Class D34 'Glen' 4-4-0s, No. 62489 *Glen Dessary*, transferred to Aberdeen Kittybrewster from Glasgow Eastfield depot to replace withdrawn ex-Great North of Scotland Railway Class D40 and D41 4-4-0s, poses in the shed yard at Keith on 4th October 1956. Built at Cowlairs Works in 1920, *Glen Dessary* was withdrawn from service in December 1959.

Keith depot

Fitted with side-window cab for Scottish Region working, Class K2/2 2-6-0 No. 61783 *Loch Sheil*, of Gresley 1921 vintage, stands on the turntable at Keith depot in October 1956. The 2-6-0 survived in service until June 1959. Its nameplates were wrongly cast by the LNER, which should have shown the correct spelling as *Loch Shiel*.

In immaculate condition, one of Keith shed's 1954 allocation of twelve ex-Great North of Scotland Railway Class D40 4-4-0s, No. 62265, built at Inverurie Works, Aberdeenshire in 1909, gleams in the morning sunshine at Keith Junction in 1956. The 4-4-0 was withdrawn from service in December of that year.

Also looking resplendent in British Railways mixed traffic engine livery of black, with red, cream and grey lining, ex-Caledonian Railway Class 3P 4-4-0 No. 54473, a Forres-based engine, stands in the yards at Keith Junction in 1956. Built at the Atlas Works of the North British Locomotive Company, Glasgow in 1916, the 4-4-0 was withdrawn from service in October 1959.

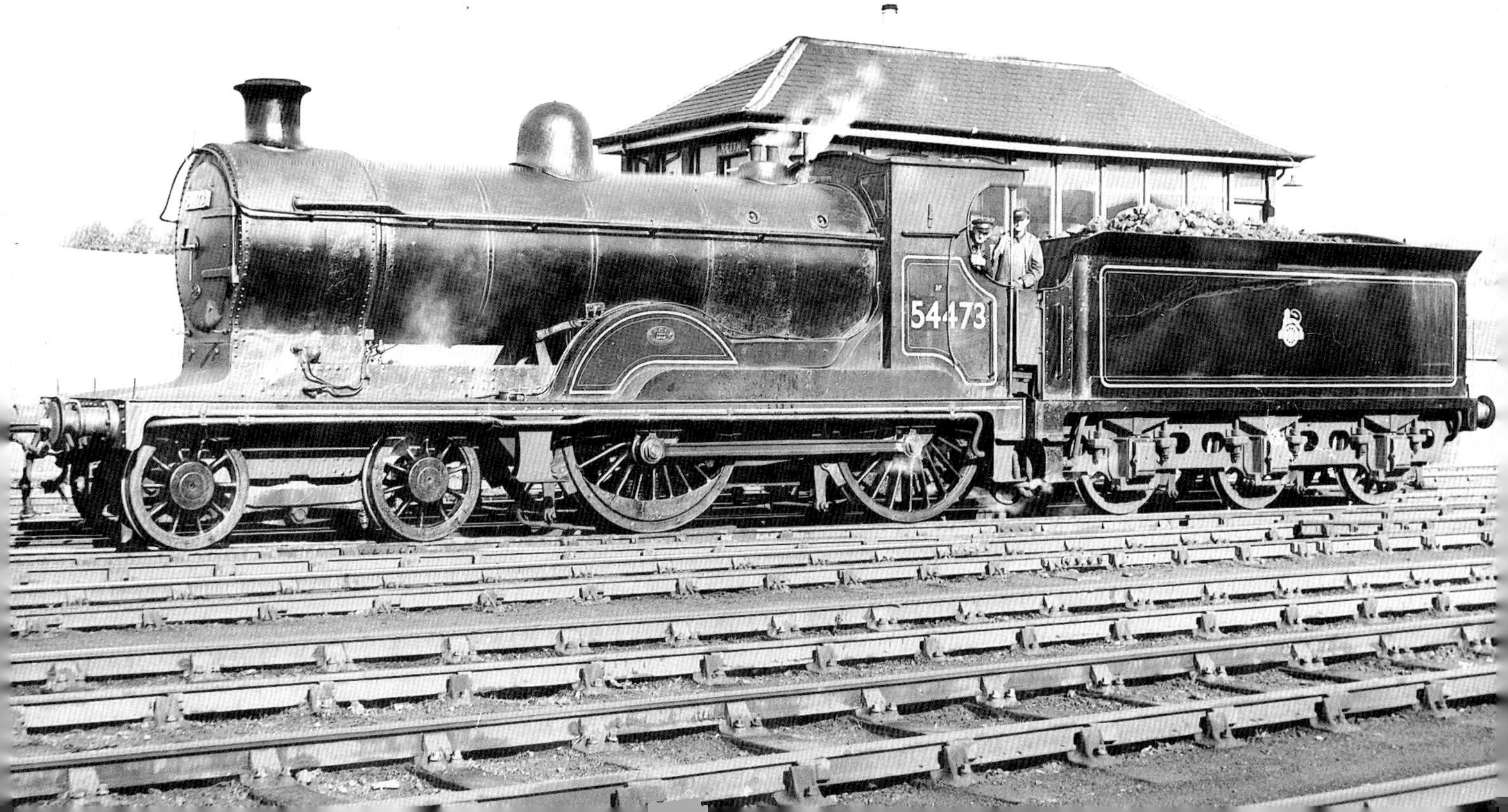

The last survivor of the ex-Great North of Scotland Railway Class D40 4-4-0s, No. 62277 *Gordon Highlander*, was built at the North British Locomotive Company Works, Glasgow in 1920. With its name painted on the splasher, the 4-4-0 is about to leave Elgin, Morayshire, with the Lossiemouth branch passenger train in October 1956. Fortunately, this attractive little engine, withdrawn from Keith shed in June 1958, was spared the scrap heap, and with brass nameplates reinstated, No. 49 was restored to her former glory in its green pre-Grouping company livery.

Gordon Highlander

As a comparison, the locomotive is shown in her final restored condition, three years later at Dumfries shed, after working a special rail tour from Glasgow. No. 49 is now on static display at the Museum of Transport, Kelvin Hall, Glasgow.

With an allocation of six ex-Caledonian Railway engines in 1954, the two-road former Highland Railway shed at Forres, Morayshire, built in 1863, is host to Class 3P 4-4-0 No. 54470, whilst under cover is Class 3F 0-6-0T No. 56301, on 7th September 1958.

Forres'shed and Cromdale

On the Speyside line, a view looking towards Craigellachie as Class 3F 0-6-0 No. 57591 pauses at Cromdale station with a passenger train to Nethy Bridge, Boat of Garten and Aviemore on 10th August 1956.

Perth, Argyll and Inverness

Alongside vintage Caledonian Railway signals, one of Perth shed's 1954 allocation of two Class D34 4-4-0s, No. 62484 *Glen Lyon*, shunts empty coaching stock at the south end of Perth General station on a summer's evening in June 1956. Built for the North British Railway in 1919, the 4-4-0 was withdrawn from Hawick shed in November 1961.

Perth General station

Class 3P 4-4-0 No. 54489 of Perth depot had been allocated the same duty when photographed on station 'pilot' work at the south end of Perth General station on 1st September 1956.

No. 123, the preserved, blue-liveried Caledonian Railway 4-2-2 of 1886 vintage, heads east on the Perth to Dundee main line alongside the River Tay, between Kinfauns and Glencarse, with an enthusiasts' special comprising the two restored Caledonian Railway coaches in the summer of 1958. As LMSR No. 14010, painted in maroon livery and allocated to Perth depot, the 4-2-2 would have been a regular performer on this line in the late 1920s.

Perth departures

British Railways Standard Class 5 4-6-0 No. 73000, the first engine of the class, pulls away from Perth General station and under the A90 road bridge at the south end of the station with the 3.30 pm Aberdeen to Glasgow Buchanan Street express on 18th September 1965.

Stanier Class 5MT 4-6-0 No. 44879, in clean condition, bustles out of Perth and past the motive power depot with a four-coach afternoon passenger train working to Gleneagles, Dunblane and Stirling on 10th May 1958.

The sun is setting on a spring evening in April 1956 as British Railways Standard Class 5 4-6-0 No. 73120 accelerates the Aberdeen portion of the southbound 'Postals' out of Perth General station.

Stored out of use after sustaining front-end collision damage, Class 4P 3-cylinder Compound 4-4-0 No. 40939 of Forfar shed, was recorded at the rear of Perth depot on 14th April 1956.

Perth shed

A scene at the busy Perth locomotive depot on a July afternoon in 1957. Awaiting the call of duty are, from left to right, Stanier Class 5MT 4-6-0 No. 45472, Class 6P 'Jubilee' 4-6-0 No. 45673 *Keppel*, and another Class 5MT 4-6-0, No. 44798. From a total allocation of 114 steam engines in 1954, no fewer than 65 Stanier Class 5MT 4-6-0s were based at Perth.

Built in 1885 at St Rollox Works, Glasgow, Drummond 'Standard Goods' Class 2F 0-6-0 No. 57345, fitted with a Westinghouse pump and stovepipe chimney, sits in the late afternoon sun at Perth shed after completing a spell of shunting duty in the adjoining yards on 2nd July 1956.

Locomotive variety at Perth

Inverness-based Stanier Class 5MT 4-6-0 No. 45360, fitted with a small bufferbeam snowplough and tablet catcher has just been turned at Perth shed in readiness to work a northbound freight on the Highland main line over Druimuachdar and Slochd summits on the 14th April 1956.

Gresley Class K4 2-6-0 No. 61996 *Lord of the Isles* of Glasgow Eastfield depot, leaves Perth shed to work a southbound evening freight in July 1957.

Amidst attractive scenery below Ben Ledi (2,873 ft), Class 5MT 4-6-0 No. 45396 heads northwards towards Lochearnhead through the wooded Pass of Leny, between Callander and Strathyre, Perthshire, with the 4.25 pm Glasgow Buchanan Street and Edinburgh Princes Street to Oban passenger train on 3rd July 1957.

Oban line

Stanier Class 5MT 4-6-0 No. 45468 drifts into Callander station with the Saturdays-only 12.40 pm Oban to Edinburgh Princes Street passenger train on a July day in 1957.

The 1.20 pm Mallaig to Fort William and Glasgow Queen Street passenger train about to leave Crianlarich Upper station at the scheduled time of 5.23 pm, behind Class 5MT 4-6-0 No. 44968 on a glorious summer's afternoon in July 1957.

West Highland line through Perthshire

A picture taken from the carriage window of the same working the following year, on 11th September 1958. The train is crossing the county boundary between Inverness-shire and Perthshire amidst typical West Highland scenery near Tyndrum, behind Class 5MT 4-6-0 No. 44968 piloting Class K2/2 No. 61789 *Loch Laidon*.

Opened in 1886, the five-mile long Caledonian Railway branch line to Killin and Loch Tay pier, Perthshire, left the Glasgow Buchanan Street to Oban main line at Killin Junction and continued on a falling gradient to the terminus at the western end of Loch Tay. The single-line branch train originally connected with steamer services on Loch Tay, but the sailings were withdrawn in 1939. The branch was closed completely in 1965.

With Ben Lawers and the Perthshire mountains as a backdrop, Class 2P 0-4-4T No. 55222 of Stirling depot heads away from Killin with the 1.52 pm branch passenger train to Killin Junction on 1st July 1957.

Killin branch

On the same July day in 1957, Class 2P 0-4-4T No. 55176 runs bunker first down the valley of the River Dochart towards Killin with the daily branch goods train.

Curving away from the Perth to Inverness main line at Ballinluig Junction, between Dunkeld and Pitlochry, Perthshire, the 8¾-mile former Highland Railway branch followed the River Tay and served two intermediate stations, at Grandtully and Balnaguard.

After picking up passengers from a main line connecting train at Ballinluig, Class 2P 0-4-4T No. 55212, the branch engine supplied by Perth shed, storms out of the junction past a dual signal controlling the single line, with the 4.55 pm passenger train to Aberfeldy on 29th June 1957.

Aberfeldy branch

The tranquil setting of Aberfeldy station terminus as 0-4-4T No. 55212 simmers at the platform before returning to Ballinluig, running bunker first with the 5.25 pm branch passenger train on 5th July 1957.

An evocative rural scene at Aberfeldy as the branch engine, 0-4-4T No. 55212, sits outside the wooden shed building, which has lost part of its roof, before working the 5.25 pm branch passenger train back to Ballinluig Junction on 5th July 1957.

The crew of Class 3P 4-4-0 No. 54499 prepare to leave the rose-adorned Ballinluig Junction station and head north alongside the River Tummel towards Pitlochry with the 10.15 am all-stations passenger train from Perth to Blair Atholl and Aviemore on 5th July 1957.

At Strathord, to the south of Stanley Junction, a short branch line connected the village of Bankfoot with the Perth to Inverness main line. Overlooked by the church at the Bankfoot station terminus, Class 4F 0-6-0 No. 44328 of Perth depot, is about to leave tender first, with the restored Caledonian Railway coaches on a Stephenson Locomotive Society special train, No. A123, on 11th October 1958. The former Caledonian Railway Bankfoot branch line was closed to passenger traffic in April 1951.

Bankfoot and Almondbank

Passing Almondbank on the ex-Caledonian Railway line to Crieff and Comrie, the restored CR 4-2-2 No. 123 looks resplendent in her sky blue livery as she heads west from Perth with the Stephenson Locomotive Society special train on the same day.

Class 2P 0-4-4T No. 55215 leaves Oban station terminus with an afternoon local branch passenger train to Connel Ferry and Ballachulish as Class 5MT 4-6-0 No. 45400 waits at the station platform with a passenger train working to Glasgow Buchanan Street on 20th July 1957.

Oban

The 6.35 pm Saturdays-only through train from Oban to Glasgow Buchanan Street conveying a restaurant and sleeping car for London Euston, raises the echoes as it blasts up the 1 in 50 gradient towards Glencruitten Summit between Oban and Connel Ferry, behind Class 5MT 4-6-0s No. 44922 piloting No. 44998 in the summer of 1959.

Dominated by the surrounding slate quarries, the station terminus at Ballachulish, Argyllshire, at the head of Glen Coe, presents a rather bleak appearance on an overcast day in September. Class 2P 0-4-4T No. 55263 is in charge of the Oban branch passenger train.

Ballachulish branch

On the Ballachulish branch single line near Kentallen on the southern shores of Loch Linnhe, Class 3F 0-6-0 No. 57667 leaves the station and ambles along the branch with the afternoon goods train for Oban via Connel Ferry in the summer of 1960.

On the 27-mile long Ballachulish branch line, a clean Class 2MT 2-6-0, No. 46468, recently transferred to Oban shed, follows the shore line of Loch Linnhe near Appin, Argyllshire, with a three-coach passenger train from Oban and Connel Ferry, the main line junction, in the summer of 1960. After crossing the two-span cantilever viaduct over the narrows at Loch Etive, the former Caledonian Railway branch line served stations at North Connel, Benderloch, Barcaldine Halt, Creagan, Appin, Duror, Kentallen, Ballachulish Ferry and Ballachulish (Glen Coe). Closure of the branch line took place in March 1966.

On the 18th June 1960, a steam-hauled return special train was run over the West Highland line from Glasgow Queen Street to Fort William. Organised by the Stephenson Locomotive Society, The 'Jacobite' was hauled throughout by Class K4 2-6-0 No. 61995 *Cameron of Lochiel*. After working the special train, No. 61995 returned to its home depot at Thornton Junction, Fife, where it remained at work until its withdrawal from service in October 1961. The K4 2-6-0 is shown being serviced at Fort William depot after the outward trip. Ex-LNER 'Coronation' observation car No. E1729E is visible on the shed turntable to the rear of the locomotive.

West Highland line special

The SLS special train of 18th June 1960 on arrival at the old Fort William station which was later demolished to accommodate a new road layout towards the town centre. The present station was built on a site further east, towards Mallaig Junction and was brought into use in 1975.

A section of the steam tour participants gathered at Bridge of Orchy station before the special sets off northwards across the bleak Rannoch Moor on the outward journey to Fort William.

The view ahead from the 1.20 pm Mallaig to Fort William and Glasgow Queen Street passenger train with fish wagons attached as Thompson 'rebuild' of 1945, Class K1/1 2-6-0 No. 61997 *MacCailin Mór* gets to grips with the sharp curves near Morar on the climb out of Mallaig on 11th June 1958.

West Highland line – Mallaig Extension

Three miles from its destination at the Mallaig terminus, the 4.40 pm passenger train from Fort William and Glasgow Queen Street leaves Morar behind Class K1/1 2-6-0 No. 61997 *MacCailin Mór* in September 1958.

As part of a joint Stephenson Locomotive Society and Railway Correspondence & Travel Society week-long Scottish rail tour on 15th June 1960, the veteran Highland Railway 'Jones Goods' 4-6-0 No. 103 worked over the 118-mile former Highland Railway main line between Perth and Inverness with a five-coach formation including the two restored Caledonian Railway coaches. No. 103 is seen heading the special train along the Spey Valley between Aviemore and Kincraig on the return journey from Inverness.

Highland Railway 4-6-0 No. 103

Two years later, on 21st April 1962, No. 103 was recorded at Gollanfield station between Inverness and Nairn, the junction for the 1-mile long former Highland Railway branch line to Fort George at the western end of the Moray Firth. The engine was heading a Stephenson Locomotive Society and Branch Line Society 'Scottish Rambler' tour train.

Highland Railway No. 103, the first 4-6-0 locomotive type to be built for use in Great Britain, waits to leave Inverness station with the Stephenson Locomotive Society and the Railway Correspondence & Travel Society return special train to Perth on 15th June 1960. Built by Sharp, Stewart & Company at the Atlas Works, Glasgow in September 1894, No. 103 became LMSR No. 17916 and was withdrawn from service in 1934. At a time when steam traction was taken for granted, it was indeed fortunate that this historic locomotive with its distinctive fluted chimney, was spared the scrap heap.

En route for Perth on the return southbound journey from Inverness with the special train, Highland Railway No. 103, in its yellow ochre livery, pauses at Aviemore station to take on water. The 'Jones Goods' 4-6-0 is now on static display at the Museum of Transport, Kelvin Hall, Glasgow.

Ross & Cromarty, Sutherland and Caithness

The foothills of the 2,396-ft high Beinn na Caillich on the Isle of Skye dominates the scene across the narrow Kyle of Lochalsh to Kyleakin as Class 5MT 4-6-0 No. 45497, recently ex-works after overhaul, waits to leave the station terminus with the 10.45 am all-stations to Dingwall and Inverness passenger train on 9th September 1958.

Kyle of Lochalsh – terminus for Skye

On the same day, the regular Kyle of Lochalsh station 'pilot' engine, Class 2P 0-4-4T No. 55216 of Inverness depot, sub-shedded at Kyle, takes a short break during shunting operations at the terminus.

With the mountains of the Isle of Skye in the distance, Stanier Class 5MT 4-6-0 No. 45463 skirts the southern shores of Loch Carron, between Kyle of Lochalsh and Duirinish, Ross and Cromarty, with the 10.45 am Kyle to Inverness passenger train via Dingwall in the summer of 1960. The Kyle line served stations at Duirinish, Plockton, Stromeferry. Attadale, Strathcarron, Achnashellach, Glencarron, Achnasheen, Achanalt, Lochuichart, Garve and Achterneed.

Driver McLean and his fireman pose beside their Stanier Class 5MT 4-6-0 at Kyle of Lochalsh terminus before taking charge of the 5.30 pm all-stations passenger train to Dingwall and Inverness.

Tain

A scene at the attractive Tain station, situated on the southern shores of the Dornoch Firth, as Class 3P 4-4-0 No. 54463 marshalls coaching stock for the 3.45 pm all-stations passenger working to Dingwall and Inverness in September 1956. Built at St Rollox in 1916, No. 54463 was withdrawn from service in December 1962 – the last survivor of the class.

No. 54463, an Inverness-based 4-4-0 is seen in action again at the south end of Dingwall station as she departs with the 1.47 pm all-stations stopping passenger via Muir of Ord and Beauly train to Inverness on 9th September 1958.

Dingwall

A spotless Class 3P Pickersgill-designed 4-4-0, No. 54487 of Inverness depot, sub-shedded at Dingwall, shunts wagons in Dingwall station yards in September 1958. One of the ten Armstrong Whitworth batch of engines built in 1921, the 'Caley Bogie' was withdrawn from service in March 1961.

The railway terminus at Thurso, Caithness, the most northerly station in the United Kingdom. On 23rd September 1959, Class 3P 4-4-0 No. 54491 prepares to leave with a passenger train for Georgemas Junction, the diverging point for lines to Thurso and Wick on the northern section of the former Highland Railway.

The Far North – Thurso and Wick

A general view of the signal box and station yards at Wick on the same day. Class 3P 4-4-0 No. 54496 shunts coaching stock at the station platform.

British Railways Scottish Region
Locomotive Depots and Sub-Sheds, April 1954

Code	Depot	Steam locomotive allocation	Sub-sheds
60A	Inverness	55	Dingwall, Kyle of Lochalsh
60B	Aviemore	8	Boat of Garten
60C	Helmsdale	5	Dornoch, Tain
60D	Wick	4	Thurso
60E	Forres	6	
61A	Aberdeen (Kittybrewster)	54	Ballater, Fraserburgh, Peterhead
61B	Aberdeen (Ferryhill)	36	
61C	Keith	19	Banff, Elgin
62A	Thornton Junction	107	Anstruther, Burntisland, Ladybank, Methil Dock
62B	Dundee (Tay Bridge)	91	Arbroath, Dundee West, Montrose, St Andrews
62C	Dunfermline (Upper)	69	Alloa
63A	Perth	114	Aberfeldy, Blair Atholl, Crieff
63B	Stirling	43	Killin, Stirling (Shore Road)
63C	Forfar	15	
63D	Fort William	12	Mallaig
63E	Oban	6	Ballachulish
64A	Edinburgh (St Margaret's)	221	Dunbar, Galashiels, Granton, North Berwick, Peebles, Seafield, South Leith
64B	Edinburgh (Haymarket)	50	
64C	Edinburgh (Dalry Road)	44	
64D	Carstairs	39	
64E	Polmont	42	
64F	Bathgate	39	
64G	Hawick	22	Kelso, Riccarton Junction, St. Boswells
65A	Glasgow (Eastfield)	139	Balloch (65I)
65B	St Rollox	74	
65C	Parkhead	66	Helensburgh (65H)
65D	Dawsholm	50	Dumbarton
65E	Kipps (Coatbridge)	49	
65F	Grangemouth	33	
65C	Yoker	13	
65H	Helensburgh	10	Arrochar
65I	Balloch	3	
66A	Glasgow (Polmadie)	178	
66B	Motherwell	104	
66C	Hamilton	52	
66D	Greenock (Ladyburn)	45	Greenock (Princes Pier)
67A	Corkerhill	92	
67B	Hurlford	58	Beith, Muirkirk
67C	Ayr	58	
67D	Ardrossan	38	
68A	Carlisle (Kingmoor)	148	Durran Hill
68B	Dumfries	38	Kirkcudbright
68C	Stranraer	13	Newton Stewart
68D	Beattock	12	
68E	Carlisle (Canal)	53	

Notes for Scottish Locomotive Surveys 1 and 2

The Locomotive Surveys 1 and 2 illustrate and briefly describe all steam engine classes officially allocated to BR Scottish Region depots in April 1954.

Scottish Locomotive Survey 1 reviews in class type, alphabetical order and by wheel arrangement, locomotives of the former LNER and its pre-Grouping classes.

Scottish Locomotive Survey 2 reviews in engine number running order, locomotives of the former LMSR and its pre-Grouping classes, British Railways Standard classes and ex-War Department engines.

The British Railways Standard Class 7MT 4-6-2 'Britannia' and the Standard Class 3MT 2-6-0 locomotives of the 77000 and 78000 series were not yet allocated to the Scottish Region during the period of the 1954 survey, but these types have, nevertheless, been included in Survey 2.

Although withdrawn from service prior to the survey, illustrations of the ex-North British Railway Class D33 4-4-0s and the ex-Great North of Scotland Class D41 4-4-0s have been included in Survey 1. Similarly, the ex-LMSR Class 4P 60 class 4-6-0 and the ex-CR Class 4P 'Wemyss Bay' 4-6-2T have been included in Survey 2 as being of particular interest, and examples of these classes may well have still been intact, although non-operational, during 1954.

Excluded from Survey 1 is the Class Y1 Sentinel 0-4-0T No. 68138 of 1927, sub-shedded at Kelso and Class G5 0-4-4T No. 67327 withdrawn from Aberdeen (Kittybrewster) in late 1954. Excluded from Survey 2 are the Fowler Class 3P 2-6-2Ts Nos 40021 and 40030 which were temporarily based at Beattock shed (68D) for banking duties, and Sentinel 0-4-0T No. 47182, built by the LMSR in 1930 and allocated to Ayr (67C).

Although the Carlisle depots of Kingmoor (68A), with a steam allocation of 148 locomotives and Canal (68E) with a steam allocation of 53 locomotives were included within British Railways Scottish Region for administrative purposes, they fall outside the geographical limits of this book.

Scottish Locomotive Survey 1

LNER and constituent company classes allocated to Scottish Region depots in April 1954.

Class A1 4-6-2 No. 60152 *Holyrood*. The Class A1 4-6-2s were a development of the Class A1/1 Pacific No. 60113 (LNER No. 4470) *Great Northern*, a Thompson rebuild of 1945. From a class total of 49 locomotives, BR Nos 60114–60162, built at Darlington and Doncaster in 1948/9 for express passenger work, five Class A1s were allocated to Edinburgh Haymarket depot in 1954: Nos 60152 *Holyrood*, 60159 *Bonnie Dundee*, 60160 *Auld Reekie*, 60161 *North British* and 60162 *Saint Johnstoun*. No. 60152 is seen at Haymarket Central Junction, Edinburgh on 8th September 1957. Named in 1951, the Pacific was withdrawn from service in June 1965.

Ex-LNER Class A2 4-6-2 No. 60528 *Tudor Minstrel*.

From a Peppercorn design of 15 Pacific locomotives built at Doncaster Works in 1947/8, BR Nos 60525–60539, a total of eleven Class A2 4-6-2s were allocated to the Scottish Region: Nos 60525/7/8/9/30/1/2/4/5/6/7, and were based at Edinburgh Haymarket, Aberdeen Ferryhill and Dundee Tay Bridge depots. No. 60539 *Bronzino*, built in August 1948, was fitted with a double chimney and Nos 60526/9/32/3/8 were rebuilt with double chimneys in 1949. No. 60528, a Dundee Tay Bridge depot-based engine, was photographed at Dalmeny on 14th April 1956. The engine was withdrawn in June 1966. No. 60532 *Blue Peter* is preserved in working order.

Ex-LNER Class A2/1 4-6-2 No. 60509 *Waverley*.

A class of four Thompson-designed locomotives, BR Nos 60507–60510, were built at Darlington Works in 1944 with Class V2 type boilers. In 1954, three engines were based at Edinburgh Haymarket: Nos 60507 *Highland Chieftain*, 60509 *Waverley* and 60510 *Robert the Bruce*. No. 60508 *Duke of Rothesay*, was allocated to Peterborough depot and withdrawn from service in February 1961. One Class A2/3 4-6-2, No. 60519 *Honeyway*, built in 1947 and also allocated to Haymarket, survived in service until December 1962.

Ex-LNER Class A3 4-6-2 No. 60099 *Call Boy*.
A world famous Sir Nigel Gresley development of the Great Northern Railway Class A1 Pacific locomotives Nos 1470 and 1471 built in 1922 and designed to haul the fastest and heaviest trains on the LNER. From a class total of 78 engines, BR Nos 60035–60112, 15 of the A3s, Nos 60035/37/41/3/57/87/9/90/4/6–9/100/1, were Scottish-based in 1954 and allocated to Edinburgh Haymarket depot. A further four of the class, Nos 60068/79/93/5 were based at Carlisle Canal depot for Waverley route working. No. 60099, originally LNER No. 2795, was built at Doncaster in 1930 and was photographed at Dalmeny station in April 1956. *Call Boy* remained in service until October 1963.

Ex-LNER Class A4 4-6-2 No. 60012 *Commonwealth of Australia*.
A celebrated class of 34 streamlined locomotives, BR Nos 60001–60034, designed by Sir Nigel Gresley and introduced in September 1935 for high speed East Coast Main Line passenger services between London King's Cross, Newcastle and Edinburgh Waverley. Seven of the Class A4s, Nos 60004/9/11/2/24/7/31, were based at Edinburgh Haymarket in the 1950s and these locomotives performed to the highest standards of steam traction on the 393.7-mile long 'non-stop' Anglo-Scottish ECML schedules, hauling such expresses as the 'Capitals Limited' and the 'Elizabethan'. No. 60012, one of Haymarket shed's 'top-link' locomotives, is seen leaving the depot to work the southbound 'Elizabethan' in September 1957. This A4 was withdrawn from service seven years later, in August 1964.

Ex-LNER Class B1 4-6-0 No. 61007 *Klipspringer*.
A Thompson 'Utility' design built between 1942 and 1947, 410 locomotives of the class were constructed at the North British Locomotive Company, Glasgow, the Vulcan Foundry, Gorton and Darlington works. The B1 4-6-0s, BR Nos 61000–61409, were widely distributed throughout Scotland on mixed traffic duties. Forty engines of the class were named after species of South African antelope and 18 engines received the names of directors of the former LNER. No. 61007 was recorded at Haymarket in June 1957.

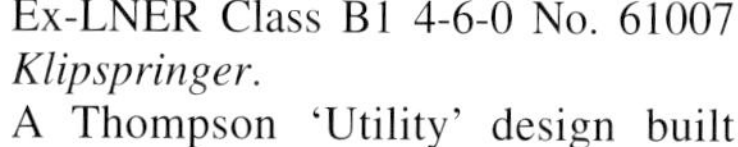

Ex-GER Class B12/1 4-6-0 No. 61502. An S. Holden design of 1911 built for the Great Eastern Railway, around twelve of these engines were based at Aberdeen Kittybrewster and Keith depots. A few locomotives of the class allocated to the former Great North of Scotland territory retained their LNER apple green livery until withdrawal from service in the mid-1950s. No. 61502 is shown at Keith in 1953. No. 61539 was the last Class B12/1 to run on the GNS section.

Ex-NBR Class C15 4-4-2T No. 67459. A. W. P. Reid design of 1911, this class of 30 engines, built by the Yorkshire Engine Company, Sheffield between 1911 and 1913 was intended for use on branch line and suburban passenger services on the North British Railway system. Numbered 67452–67481 by British Railways from 1948, one of the last survivors, No. 67459 formerly LNER No. 9134, is illustrated at Polmont in October 1955 shortly before being hauled away for scrap.

Ex-NBR Class C16 4-4-2T No. 67502. A later W. P. Reid design of 1915, the Class C16s built by the North British Locomotive Company, comprised a total of 21 engines, BR Nos 67482–67502. Similar to the Class C15s, but with superheaters, the 4-4-2s were used on suburban and branch line passenger workings well into the British Railways era over former North British Railway territory. No. 67502, built in March 1921 and the last of the class numerically, survived in service until April 1960 and is seen at Dundee Tay Bridge shed in April 1956.

Ex-LNER Class D11/2 4-4-0 No. 62685 *Malcolm Graeme*.
Designed by J. G. Robinson of the Great Central Railway in 1913, the D11/2s were a later development of 24 Gresley 'improved' engines built by Kitson and Armstrong Whitworth in 1924. Modified for the Scottish loading gauge, the 'Scottish Director' 4-4-0s later became BR Nos 62671–62694 and were named after characters from Sir Walter Scott's 'Waverley' novels. No. 62685, the last surviving locomotive of the class, was recorded at Edinburgh Haymarket depot after working a Glasgow Queen Street to Edinburgh Waverley express in September 1957.

Ex-NBR Class D30/2 4-4-0 No. 62427 *Dumbiedykes*.
Introduced by W. P. Reid in 1914 for the North British Railway and built at Cowlairs Works, Glasgow, these 4-4-0s were known as 'Scotts'. Designed to handle local and express passenger services, these familiar engines, BR Nos 62418– 62442, were developed from the earlier NBR Class D30/1 and D33 types. A Dunfermline-based engine, No. 62427 is shown at Edinburgh Haymarket depot in May 1956.

Ex-NBR Class D33 4-4-0 No. 62464. Also based at Dunfermline, the last survivor of the twelve Reid 'Intermediate' Class D33 4-4-0s, No. 62464, was built at Cowlairs Works, Glasgow in 1909. The engine is seen at Edinburgh Haymarket shed in its final days of service after working a local passenger train from Fife to Edinburgh Waverley in February 1954. No. 62464 was later broken up at Kilmarnock Works.

Ex-NBR Class D34 4-4-0 No. 62467 *Glenfinnan*.
W. P. Reid's celebrated class of 32 engines, BR Nos 62467–62498, built at Cowlairs Works, Glasgow and introduced in 1913, were a small-wheeled version of the 'Scotts' and designed to work over the West Highland line. The 4-4-0s were named after Scottish glens. No. 62467 of Thornton Junction depot, prepares to leave Haymarket depot to head a local passenger train from Edinburgh Waverley to Fife in August 1958. No. 62469 *Glen Douglas* was restored to its original North British Railway condition as No. 256.

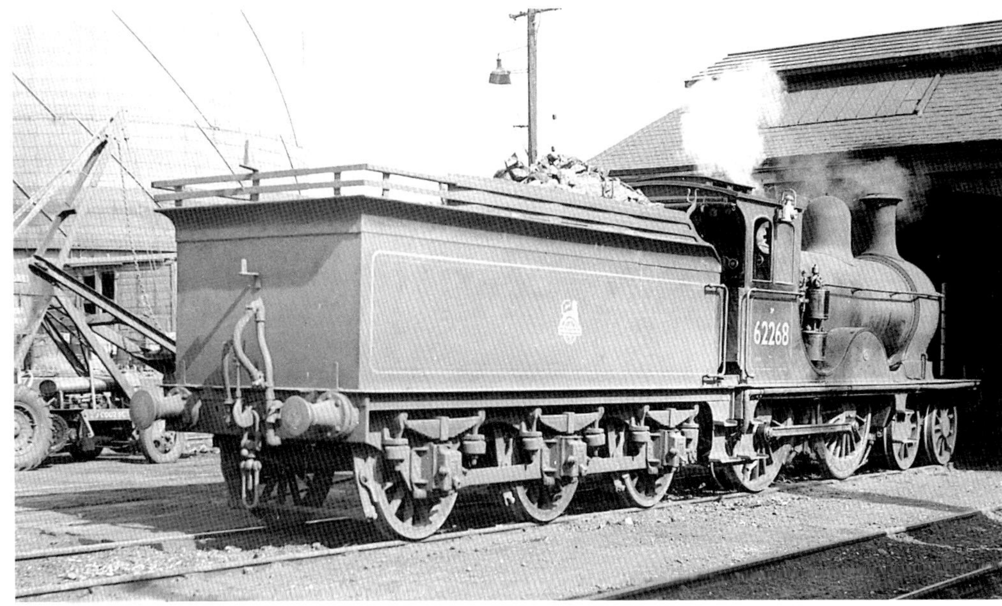

Ex-GNSR Class D40 4-4-0 No. 62268. Introduced by W. Pickersgill in 1899, a further batch of T. E. Heywood-designed engines followed in 1920. Eight of the later 4-4-0s were superheated and named after places and personalities associated with the Great North of Scotland Railway. Fifteen engines of the class were still at work in mid-1954, BR Nos 62262/4/5/7/8/9/71–5/7, 62276/8/9. D40 class No. 62277 *Gordon Highlander* was preserved and restored to GNSR livery. Inverurie Works-built No. 62268 is shown at Keith depot in 1956.

Ex-GNSR Class D41 4-4-0 No. 62242. A few examples of the Pickersgill D41 class engines of 1893 vintage, numbered 62225–62252 by British Railways, survived into the early 1950s. No. 62242, built by Neilson & Company, Glasgow in 1895 and formerly based at Aberdeen Kittybrewster shed, awaits the breaker's torch at Inverurie Works, Aberdeenshire in 1953. The Class D40 and D41 4-4-0s were confined to mixed traffic work over the former Great North of Scotland Railway territory in the north eastern area of the Scottish Region.

Ex-LNER Class D49/1 4-4-0 No. 62719 *Peebles-shire*.
Introduced by Sir Nigel Gresley in 1927, the 34 engines of the 'Shire' class, BR Nos 62700–62725 and 62728–62735, were designed for light passenger work. Built by the LNER at Darlington, 15 of the 4-4-0s were allocated to depots north of the border in the mid-1950s and received the names of Scottish shires. No. 62719 seen is awaiting repair at Edinburgh Haymarket shed in March 1959.

Ex-LNER Class D49/2 4-4-0 No. 62743 *The Cleveland*.
A development of Sir Nigel Gresley's 'Shire' class of 1927, (above) the 'Hunt' class of 1928, BR Nos 62726/7 and 62736–62775, were fitted with Lentz rotary cam poppet valves and received the names of hunting packs. Only two of the 4-4-0s were based in Scotland in the 1950s: No. 62743 at Edinburgh Haymarket depot and No. 62744 *The Holderness* at Dundee Tay Bridge. No. 62768 *The Morpeth* was a Thompson rebuild of 1942 with two inside cylinders and was classified as D49/4. No. 62743 is shown outside Haymarket depot on 6th May 1956.

Ex-GER Class F4 2-4-2T No. 67157. An 1884 Wilson Worsdell design for the Great Eastern Railway, one engine of this class, BR Nos 67151–67187, was based at Aberdeen Kittybrewster depot in 1954 for working the Fraserburgh to St Combs branch line in north east Aberdeenshire. As the line was unfenced, the Class F4 carried cow-catchers. No. 67157 was photographed at Aberdeen Ferryhill shed in 1956 and spent its final days as works shunter at Inverurie Works until its withdrawal in the same year.

Ex-NBR Class J35/4 0-6-0 No. 64518. Designed by W. Reid in 1906 for freight haulage, the Class J35 0-6-0s, BR Nos 64460–64535 were widely used over former North British Railway lines and a total of 70 engines of the class were taken into British Railways ownership. Nos 64460–64477 were classified as J35/5. No. 64518, an Edinburgh St Margaret's-based engine is seen at Craigentinny sidings, a weekend locomotive storage point alongside the East Coast Main Line near Edinburgh, in August 1956.

Ex-NBR Class J36 0-6-0 No. 65235 *Gough*.
Designed by Matthew Holmes for the North British Railway and introduced in 1888 for freight haulage, many engines of the class, BR Nos 65210–65346, were still at work in the early 1960s. Twenty-five of the 0-6-0s saw overseas service in 1917/18 and received names associated with military leaders and places connected with the First World War.

Nos 65285 and 65287 were fitted with cut-down boiler mountings and were based at Kipps shed, Coatbridge. No. 65235, one of the named engines, is shown near Haymarket West Junction, Edinburgh in May 1956. No. 65243 *Maude*, is preserved as NBR No. 673 at Bo'ness.

Ex-NBR Class J37 0-6-0 No. 64603. A superheated version of the Class J35 0-6-0 (top) introduced by W. P. Reid at Cowlairs Works, Glasgow in 1914 for freight working, this long-lived class of over 100 engines, BR Nos 64536–64639, was used extensively by the Scottish Region well into the 1960s on freight and passenger duties. No. 64603 is shown at Craigentinny, Edinburgh, in ex-works condition on 12th August 1956.

Ex-LNER Class J38 0-6-0 No. 65917. All 35 engines of this 1926 Gresley design, built at Darlington Works, BR Nos 65900–65934, were based in Scotland, their workings confined to heavy freight, coal, and mineral traffic in the Lothian and Fife areas. A freight version of the Class J39/2 (below), No. 65917, rebuilt with Class J39 boiler and recently ex-works, was photographed on the shed turntable at Polmont on 21st May 1956.

Ex-LNER Class J39/2 0-6-0 No. 64950. Also designed by Gresley, but with 5ft 2in diameter driving wheels, construction of this class followed later in 1926. From a class total of 289 engines, BR Nos 64700–64988, around 20 Class J39/2s were allocated to former LNER depots in Scotland during 1954. The 0-6-0s were used mainly for freight haulage but occasionally appeared on local passenger and excursion trains. No. 64950 is seen at her home depot of Dundee Tay Bridge on 14th April 1956. The last engine of the class, No. 64986, was withdrawn from Thornton Junction shed in December 1962.

Ex-LNER Class J50/3 0-6-0T No. 68954. Another Gresley design introduced at Doncaster in 1926, seven of these engines were allocated to Glasgow Eastfield depot in 1954 and were employed on shunting duties in the marshalling yards at Cadder, east of the city. The J50 class variations were numbered 68890–68991 by British Railways. In August 1958, No. 68954 was recorded at Eastfield shed, out of use pending withdrawal from service.

Ex-GER Class J69/1 0-6-0T No. 68503. A 1902 development of the Class J67 built for the Great Eastern Railway, only a few examples of this numerous class, BR Nos 68490–68636, were allocated to Scottish depots, at Edinburgh St Margaret's, Thornton Junction and Polmont. Holden design No. 68503 was based at Glasgow Parkhead shed and was recorded there out of use in June 1956.

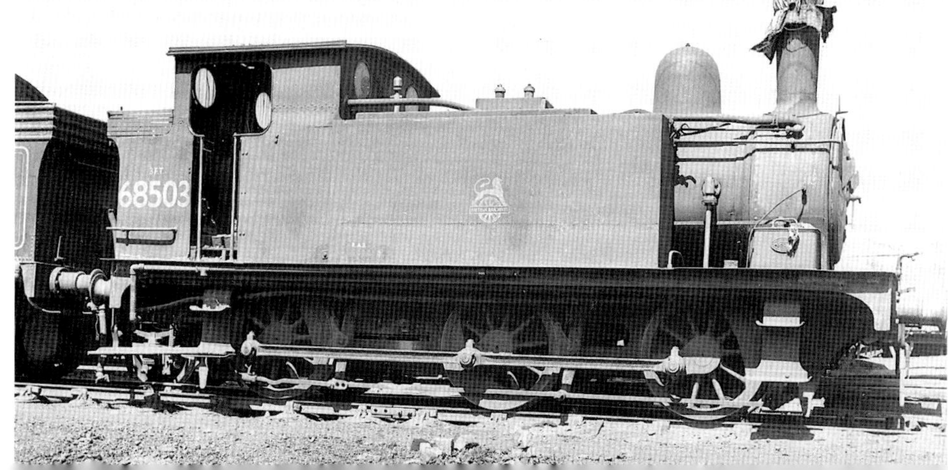

Class J72 0-6-0T No. 69014. An 1898 Wilson Worsdell design for the North Eastern Railway, BR Nos 68670– 68754, a further construction of 28 engines, BR Nos 69001–69028, commenced as late as 1950 to identical specifications. The 1954 Scottish Region allocation of ten Class J72 0-6-0Ts was shared between Aberdeen Kittybrewster, Edinburgh St Margaret's, Glasgow Eastfield, and Thornton Junction, Fife depots. No. 69014, an example of the later batch of 0-6-0Ts, a St Margaret's-based engine, was photographed at South Leith, Edinburgh in April 1956. No. 69023, built at Darlington in 1951, has been preserved.

Ex-NBR Class J83 0-6-0T No. 68454. Introduced for light freight and station pilot duties by Matthew Holmes in 1900, and built by Messrs Neilson, Reid and Sharp, Stewart & Company, this class of 39 engines, BR Nos 68442–68481, was allocated to former North British Railway depots in the Edinburgh, Fife and Glasgow areas. Fourteen engines of the class were fitted with continuous brake. No. 68454 was recorded at South Leith, Edinburgh in ex-Cowlairs Works condition in April 1956.

Ex-NBR Class J88 0-6-0T No. 68339. Designed by W. P. Reid, with short wheelbase for sharp curve and light freight shunting work, the J88 0-6-0Ts built between 1905 and 1919, were allocated throughout the former North British Railway system, chiefly in the Edinburgh and Glasgow areas. All engines of the class, BR Nos 68320– 68354, survived into British Railways ownership and five of the 0-6-0Ts were fitted with spark arresters. No. 68339, fitted with 1909 boiler and safety valves, was the Haymarket shed 'pilot' and was photographed at the depot in May 1956.

Class K1 2-6-0 No. 62031.
Not introduced into service until after Nationalisation in 1948, a class of 70 mixed traffic locomotives, BR Nos 62001–62070, was built by the North British Locomotive Company, Glasgow in 1949/50. The 2-6-0s were a Peppercorn development of the Thompson-designed 2-cylinder Class K1/1, No. 61997 *MacCailin Mór* of 1945. In 1954, five engines of the class were based in Scotland for West Highland line work – three at Glasgow Eastfield and two at Fort William depots. Eastfield shed's No. 62031, is shown hard at work near Glenfinnan with a Mallaig bound freight on 13th August 1957. No. 62005 has been preserved in working order in LNER green livery.

(Courtesy, Hugh Ramsey)

Ex-GNR Class K2/2 2-6-0 No. 61776.
Carrying BR Nos 61721–61794, the class K2 2-6-0s were well-known in Scotland, especially on the West Highland line. Thirty engines of this 1914 Gresley design were fitted with side-window cabs and in 1933/4, 13 of the class were named after Scottish lochs. In June 1956, No. 61776, an unnamed example built by Kitson & Company of Leeds in 1921, was recorded at Glasgow Eastfield depot which, in 1954, shared the allocation of the 'Lochs' with Fort William. No. 61776 was withdrawn from service in March 1959.

Ex-LNER Class K3/2 2-6-0 No. 61823.
A 1924 development of a Nigel Gresley design of 1920 for the Great Northern Railway, all 193 members of the class, BR Nos 61800–61992, were taken into British Railways ownership including the Thompson rebuild, No. 61863 which was classified as K5. The Scottish allocation mainly worked freight trains on the East Coast and Aberdeen main lines and over the Waverley route to Carlisle. Darlington-built No. 61823 of St Margaret's depot passes Gorgie with an eastbound Edinburgh suburban line freight on 9th June 1959.

Ex-LNER Class K4 2-6-0 No. 61995 *Cameron of Lochiel*.
Designed by Sir Nigel Gresley for West Highland line working and introduced in 1937, five engines of the class, BR Nos 61993–61996 and No. 61998, were allocated to Glasgow Eastfield in 1954 and latterly at Thornton Junction depot, Fife. No. 61997 *MacCailin Mór*, the 2-cylinder Thompson rebuild of November 1945, was based at Fort William. No. 61995 is seen at Perth shed on 14th April 1956, fitted with a small bufferbeam snowplough. Withdrawn from hornton Junction in December 1961, No. 61994 *The Great Marquess* has been restored to its original LNER condition as No. 3442 and is shown on the dust jacket in its British Railways'condition.

Ex-GNR Class N2/3 0-6-2T No. 69565.
Of Great Northern Railway origin and designed by Gresley in 1925 for suburban passenger work, most of the Class N2 0-6-2Ts, BR Nos 69490–69596, were based in the London King's Cross area. The Class N2/3s were built without condensers for suburban services around Glasgow and in the early 1950s, ten of the 0-6-2Ts were allocated to Glasgow Parkhead, three were based at Kipps Coatbridge and three at Glasgow Dawsholm shed. No. 69565 lies out of use at Parkhead depot on 10th June 1956.

Ex-NBR Class N15/1 0-6-2T No. 69173.
A 60-ton Reid design of 1910 which was developed from the earlier Class N14 engines – the Cowlairs Incline banking locomotives – and carrying BR Nos 69126–69131 were classified as Class N15/2. Slightly lighter engines, with smaller bunkers, (Class N15/1). BR Nos 69132–69224, were distributed widely throughout the former North British Railway system on shunting and pilot duties. Ex-works No. 69173 was photographed at South Leith, Edinburgh on 1st April 1956. The last survivor of the class, No. 69178, was withdrawn from Motherwell shed in December 1962, and is illustrated on page 54.

Ex-LNER Class Q1/1 0-8-0T No. 69925.

Four engines of this class were Thompson rebuilds of the Robinson-designed Class Q4 0-8-0 of 1902. Introduced by the LNER in 1942/3, the Class Q1 and Q1/1 tank engines, BR Nos 69925–69937, totalled 13 locomotives, two of which were allocated to Glasgow Eastfield. The engines were employed on heavy shunting duties in the freight marshalling yards at Cadder, near Glasgow. No. 69925, with 1,500 gallon capacity tanks, is shown at Eastfield shed in 1955.

Ex-LNER Class V1 2-6-2T No. 67610. A Gresley designed locomotive introduced from Doncaster Works in September 1930 for suburban passenger work, the 2-6-2Ts were a class of 92 engines, BR Nos 67600–67691, including the later modified design Class V3s of 1939. The Scottish allocation was widely used around Edinburgh, Glasgow and in the Fife area. No. 67610, an Edinburgh Haymarket-based engine, is seen here at its home depot in May 1956.

Ex-LNER Class V2 2-6-2 No. 60873 *Coldstreamer*.

This famous Sir Nigel Gresley-designed class of 184 engines, BR Nos 60800–60983, first appeared in June 1936 and subsequently performed remarkable work, especially during the war years. The first engine of the class, LNER No. 4800, was named *Green Arrow* and a further seven engines received the names of regiments and schools in the North East of England. In the British Railways era, Scottish-based Class V2s worked passenger and express freight trains over former LNER east-coast lines, the Waverley route, and fish and meat traffic from Aberdeen. No. 4800 is preserved in LNER livery in the National Collection. No. 60873 is seen here, outside Edinburgh Haymarket depot in August 1958.

Ex-LNER Class V4 2-6-2 No. 61701. A class of only two Gresley lightweight engines – his final design – introduced in 1941 for use on the steeply graded West Highland line between Glasgow and Fort William. No. 61700 was named *Bantam Cock* and her sister engine, No. 61701, unofficially named 'Bantam Hen', was photographed at Edinburgh Haymarket depot in September 1954. Based at Glasgow Eastfield in 1954, both engines ended their days at Aberdeen Ferryhill depot and were withdrawn from service in 1957 and broken up at Kilmarnock.

Ex-NBR Class Y9 0-4-0ST No. 68095. Designed by Matthew Holmes for the North British Railway, the Y9s were introduced in 1882 for light shunting and dockyard work. A diminutive engine of 27 tons with dumb buffers, some engines of the class were permanently attached to a 6-ton wooden tender containing additional coal supplies. BR Nos 68092–68124 were based mainly in the Edinburgh and Dundee areas and in 1954 five were allocated to Kipps depot, Coatbridge. No. 68095 was recorded at Seafield, Edinburgh on 25th April 1956 and this engine has since been preserved.

Ex-GNSR Class Z5 0-4-2T No. 68193.
Designed by Manning, Wardle of Leeds in 1915 for dock shunting duties in the Aberdeen area, the two Class Z5 engines, BR Nos 68192 and 68193, and the lighter Class Z4s, Nos 68190 and 68191 were allocated to Aberdeen Kittybrewster depot in 1954. No. 68193, still lettered 'British Railways', is seen at work at Aberdeen docks in the early 1950s.

Scottish Locomotive Survey 2

LMSR and constituent company classes, BR Standard classes and ex-War Department locomotives allocated to Scottish Region depots in April 1954.

Ex-LMSR Class 3MT 2-6-2T No. 40158.
A William Stanier design of 1935, developed from a Hughes parallel boiler design of 1930, 19 of these taper boiler mixed traffic engines, BR Nos 40071–40209, were allocated to Scottish Region depots in 1954. Nine members of the class were based at Glasgow Dawsholm depot and two engines were shedded at Dumfries for working the Kirkcudbright branch line trains. Five engines of the class were rebuilt with larger boilers in 1941. No. 40158, a Derby-built engine of 1937, was recorded at Dawsholm shed on 17th August 1958.

Ex-LMSR Class 2P 4-4-0 No. 40613.
A post-Grouping development of a Midland Railway design by H. Fowler introduced in 1928 for light passenger duties, most of the Scottish Region-based 4-4-0s from a class total of 136 engines, BR Nos 40563–40700, were allocated to South West area depots for working the intensive local passenger services to and from Glasgow St Enoch and throughout Ayrshire. No. 40613, a Carlisle Kingmoor-based engine was an unusual visitor to Carstairs Junction when photographed on 16th September 1956. The last Scottish engine of the class, No. 40670 was withdrawn in December 1962.

Ex-LMSR Class 4P 4-4-0 No. 40909.
First introduced at Derby in 1905 as a development by Deeley of the Johnson Midland Compound 4-4-0s, further rebuilds followed in 1914 and together with the remaining engines introduced by the LMS in 1924, the class comprised a total of 232 locomotives, BR Nos 40900–41199, (with gaps due to scrapping). The Scottish based 4-4-0s performed splendidly on passenger trains over the former Caledonian and Glasgow & South Western Railway systems of British Railways. Around 40 Compounds were allocated to Scottish Region depots in 1954, the last survivor being No. 40920 at Stranraer. No. 40909 is shown at its home depot of Glasgow Corkerhill in the summer of 1956. In 1959, No. 1000 was fully restored to its rebuilt 1914 condition in Midland Railway crimson lake livery.

Ex-LMSR Class 4MT 2-6-4T No. 42215.
A Charles Fairburn development of a Stanier design, these useful tank engines were widely distributed throughout Scotland and were used on local passenger and light freight operations. No. 42215, built by the LMSR at Derby in 1945, was one of four of the class allocated to Beattock shed in 1954 for banking purposes on Beattock Incline. The 2-6-4T is seen at Beattock Summit after providing rear-end assistance to a northbound freight train on 4th July 1959.

Ex-LMSR Class 5MT 2-6-0 No. 42802.
Introduced by George Hughes in 1926, under Sir Henry Fowler's directions, the class totalled 245 engines, BR Nos 42700–42944, 67 of which were allocated to the Scottish Region in mid-1954 for use on heavy freight haulage on the West Coast Main Line, the Midland route, and throughout the Ayrshire coalfields. Five engines of the class were fitted with non-standard valve gear. Horwich-built No. 42802, one of Carlisle Kingmoor shed's allocation of 28 'Crabs', is shown at Grangemouth in April 1957.

Ex-LMSR Class 4MT 2-6-0 No. 43138.
Designed by H. Ivatt and introduced by the LMSR in December 1947 at Horwich Works near Bolton, Lancs, the majority of this 2-6-0 taper boiler class of 162 engines, BR Nos 43000–43161 (built at Doncaster and Darlington), were based at depots south of the border. The first 50 2-6-0s were built with double chimneys, later to be replaced by the single variety. In 1954, ten of the engines were allocated to the Scottish Region including six at Glasgow Eastfield and two at Polmont. No. 43138 of Bathgate shed was recorded at Polmont on 7th June 1959.

Ex-LMSR Class 4F 0-6-0 No. 44328. Introduced by H. Fowler in 1911 for the Midland Railway, building of the 0-6-0s was continued by the LMSR in 1924 and by 1941, a total of 772 engines were in service, from eight different railway workshops. Five of the class were built for the Somerset & Dorset Joint Railway in 1922 and taken into LMSR stock in 1930. Fifty-four of the 0-6-0s were allocated to the Scottish Region in 1954, mostly at Perth and Carlisle Kingmoor depots. No. 44328 built at St Rollox, is shown at Perth in June 1956, fitted with tender cab and tablet catching apparatus for Highland line section working.

Ex-LMSR Class 5MT 4-6-0 No. 44798. One of the most successful steam locomotive classes ever built, the Stanier 'Black 5s' first appeared in 1934 for the LMSR. Up to 1951, a total of 842 engines, BR Nos 44658-45499, were put into mixed traffic service from five separate locomotive works. Ubiquitous throughout Scotland, only four of the class received names – Nos 45154 *Lanarkshire Yeomanry*, 45156 *Ayrshire Yeomanry*, 45157 *The Glasgow Highlander* and 45158 *Glasgow Yeomanry*, built in 1935 at the Armstrong Whitworth Works and based at Glasgow St Rollox depot in 1954. From 1947, various modifications were incorporated in the newer engines including the fitting of Caprotti valve gear, double chimneys and roller bearings. No. 44767 received Stephenson link motion and this engine is now preserved. No. 44798, a Horwich-built engine of 1947, is at Perth on 2nd June 1956, still carrying the 'British Railways' lettering on its tender.

Ex-LMSR Class 6P 'Jubilee' 4-6-0 No. 45715 *Invincible*.
A Stanier taper boiler development of the 'Patriot' class introduced at Crewe Works in May 1934 for express passenger work, the first engine of the class, No. 5552, being christened *Silver Jubilee*. The remainder of the class, built at Crewe, Derby and the North British Locomotive Works, Glasgow, were named after British colonies, provinces, admirals, ships, naval battles and Irish counties. Nos 45735 and 45736 were rebuilt in 1942 with larger boilers and double chimneys. Around 30 'Jubilees' from an original class total of 191 engines, BR Nos 45552–45742, were allocated to Scottish Region depots in 1954, mainly at Carlisle Kingmoor shed. No. 45715 built at Crewe in 1936, is shown speeding past Elvanfoot with a down express in April 1959.

Ex-LMSR Class 7P 'Royal Scot' 4-6-0 No. 46107 *Argyll and Sutherland Highlander*. These famous 3-cylinder locomotives first appeared in 1927 to a parallel boiler design by H. Fowler for LMSR Anglo-Scottish express services on the West Coast Main Line. The initial success of the first 50 engines, built by the North British Locomotive Company, Glasgow, led to the construction of a further 20 'Scots' at Derby in 1930. The experimental super pressure boiler engine, No. 6399 *Fury* was rebuilt in 1935 and became No. 6170 *British Legion*. From a class of 71 engines, BR Nos 46100–46170, only five of the class were Scottish-based in 1954 and allocated to Glasgow Polmadie: Nos 46102 *Black Watch*, 46104 *Scottish Borderer*, 46105 *Cameron Highlander*, 46107 *Argyll and Sutherland Highlander*, and 46121 *Highland Light Infantry, The City of Glasgow Regiment*. No. 46107, shown at her home depot in 1956, was withdrawn from service in 1962.

Ex-LMSR Class 8P 'Duchess' 4-6-2 No. 46221 *Queen Elizabeth*. Introduced at Crewe Works in 1937 by Sir William Stanier for working the newly inaugurated high-speed 'Coronation Scot' services between London Euston and Glasgow Central, these handsome 'Pacifics were an enlargement of the 'Princess Royal' class of 1933. Streamlined and in blue and silver livery when built, the first engines, Nos 6220–6224 were followed in 1938 by Nos 6225–6229 painted in LMSR maroon livery. Streamlined Nos 6235–6248 were completed at Crewe between 1939 and 1943. Nos 6230–6234 of 1938 and 6249–6257 built 1944–1948 were non-streamlined. All were eventually de-streamlined and at Nationalisation in 1948, they became BR Nos 46220–46257. In 1954, nine 'Duchess' 4-6-2s were based at Glasgow Polmadie to work Anglo-Scottish expresses, such as the 'Royal Scot' and the 'Caledonian': Nos 46220 *Coronation*, 46221 *Queen Elizabeth*, 46222 *Queen Mary*, 46223 *Princess Alice*, 46224 *Princess Alexandra*, 46227 *Duchess of Devonshire*, 46230 *Duchess of Buccleuch*, 46231 *Duchess of Atholl* and 46232 *Duchess of Montrose*. No. 46221 is shown on the turntable at Polmadie shed in April 1956, and this engine amassed a total mileage of 1,308,000 before its withdrawal in 1963.

Class 2MT 2-6-0 No. 46462.
An H. Ivatt design of 1946 for LMSR mixed traffic duties, BR Nos 46400–46527, five of these taper boiler 2-6-0s, Nos 46460–46464 were Scottish-based in 1954 and were allocated to Aberdeen Kittybrewster, Edinburgh St Margaret's and Dundee Tay Bridge depots. The lightweight engines based at St Margaret's were intended for working the Lauder and Humbie branch lines to the south of Edinburgh. No. 46462, a St Margaret's engine, was unusually recorded at the head of an eastbound passenger train of articulated coaching stock on the Edinburgh suburban line near Craiglockhart on 6th April 1957.

Ex-LMSR Class 2F 0-6-0T No. 47162. A class of ten outside cylinder 'Dock Tanks', BR Nos 47160–47169, with short wheelbase for dockyard shunting, was introduced by H. Fowler in December 1928. Five engines of the class were allocated to Scottish Region depots in 1954, at Greenock Ladyburn, Edinburgh Dalry Road and St Margaret's. No. 47162, fitted with spark arrester, was recorded at Granton Harbour, Edinburgh in 1956.

Ex-LMSR Class 3F 0-6-0T No. 47305. First introduced in 1924, the Class 3F tank engines were a post-Grouping development of a Midland Railway design by H. Fowler. A total of 417 'Jinties', BR Nos 47460–47681, were built for light shunting duties, including seven engines for the Somerset & Dorset Joint Railway, taken into LMSR stock in 1930. Only seven of the class were allocated to Scottish Region depots in 1954 – five at Glasgow Polmadie, one at Glasgow Corkerhill and one at Inverness. No. 47305, formerly based at Aintree depot, Liverpool, awaits the breaker's torch in the shipyards at Troon, Ayrshire in 1961.

Ex-CR Class 3P 4-4-0 No. 54446. Designed by J. F. McIntosh, the Caledonian Railway 139 class was built at St Rollox Works, Glasgow and introduced in 1910. Twenty-two of these engines survived into British Railways ownership. No. 54446, one of the last of the superheated 'Dunalastair IV' class 4-4-0s, (originally Caledonian Railway No. 118), poses for the camera during station 'pilot' duties at Carstairs Junction in August 1955, the month of its withdrawal from service and dispatch to Kilmarnock for scrapping.

Ex-CR Class 3P 4-4-0 No. 54503. One of the W. Pickersgill 72 class in ex-Inverurie Works condition, awaits its next turn of duty at Perth depot on 3rd July 1957. This class, together with its variants, the 113 and 928 class 4-4-0s, BR Nos 4461–54508, were usually referred to as 'Caley Bogies' and these versatile engines worked over Scottish Region lines well into the 1960s. Built at the Hyde Park Works of the North British Locomotive Company, Glasgow in 1922, No. 54503 was withdrawn from service in October 1959. The construction of the class was shared between St Rollox Works, Armstrong Whitworth & Company, and the North British Locomotive Company, Atlas and Hyde Park works, Glasgow.

Ex-LMSR Class 4P 4-6-0 No. 54639. The last survivor of the modified Class 60 4-6-0 of 1925 at Hamilton depot shortly before its withdrawal from service in 1953. The original class of six engines built for the Caledonian Railway was designed by W. Pickersgill in 1916. After the Grouping in 1923, the LMSR built a further 20 engines of this type with slightly larger cylinders. A few of the 'Caley 60s' survived into early Nationalisation and were based at Edinburgh Dalry Road, St Margaret's, Hamilton and Motherwell depots.

Ex-HR Class 1P 0-4-4T No. 55053. Highland Railway 25 Class No. 45, BR No. 55053, was a Peter Drummond design built at Lochgorm Works, Inverness in 1905. No. 55053, along with No. 55051, was based at Helmsdale for working the 7¾-mile long Dornoch branch line from The Mound, between Lairg and Helmsdale. No. 55053 is seen at St Rollox Works, Glasgow, newly outshopped in British Railways lined out black mixed traffic livery. The last ex-Highland Railway engine in regular service, the 0-4-4T was withdrawn in 1957, retired prematurely because of a broken driving axle. Two ex-GWR 0-6-0PTs were transferred north as replacement engines for the Dornoch branch services.

Ex-CR Class 2P 0-4-4T No. 55213. The McIntosh 439 class of 1900, together with its numerous variants including the 19 and 92 classes, were introduced between 1895 and 1925. The 0-4-4Ts were designed for suburban passenger, branch line and station 'pilot' work on the Caledonian Railway. Allocated to depots throughout the Scottish Region, the majority of the class. BR Nos 55124–55269, survived into British Railways ownership. No. 55213 of 1912 vintage, was shunting coaching stock at Perth General station when photographed on 2nd June 1956. One engine of the class, Caledonian Railway No. 419, survives in preservation in original blue livery and in working order.

Ex-LMSR Class 2P 0-4-4T No. 55261. A post-Grouping version of the Caledonian Railway 439 class, ten locomotives were built for the LMSR in 1925, at the Nasmyth, Wilson works in Manchester. A well-groomed member of the class, No. 55261, in full British Railways mixed traffic engine livery, pauses between station 'pilot' duties at Carstairs Junction on 17th September 1956.

Ex-CR Class 4P 4-6-2T No. 55359. The last of the twelve W. Pickersgill-designed Class 944 'Wemyss Bay' tank engines built by the North British Locomotive Company, Glasgow in 1917 for the Caledonian Railway. The express 4-6-2Ts, BR Nos 55350–55361 were, in their heyday, used on Clyde coast passenger trains from Glasgow Central and on local suburban and excursion workings. Ten engines of the class ended their days at Beattock on banking duties. No. 55359 is seen at Beattock shed shortly before its withdrawal in 1953. The 4-6-2T was broken up at Kilmarnock in 1954.

Ex-CR Class 0F 0-4-0ST No. 56025. The 611 class engines were introduced by J. McIntosh in 1895 and developed from a Drummond class of 1885 for dockyard shunting work. In 1954, examples of this small class of 'Caley Pugs' were based at Greenock Ladyburn, Yoker, Inverness and St Rollox Works. No. 56025, the Glasgow St Rollox works shunting engine of 1890 vintage, was painted in British Railways mixed traffic engine black livery lined out in red, cream and grey and was always kept in immaculate condition. The 0-4-0ST was photographed at St Rollox Works on 9th June 1957. Two engines of the class were allocated to the London Midland Region in the early 1950s also for use as works shunters.

Ex-CR Class 2F 0-6-0T No. 56164.
Introduced by J. McIntosh in 1911 for dockyard shunting and local freight working around Glasgow and Greenock, the Class 498 engines became known as 'Beetlecrushers'. No. 56164, built at St Rollox in 1918, was one of the 23 outside cylinder tank engines taken into British Railways ownership, BR Nos 56151–56173. Fitted with spark arrester for timber yard duties No. 56164 is seen in company with former 29 class 0-6-0T No. 56267 at Grangemouth in March 1956. The last engine of the class, No. 56159 was withdrawn from Polmadie shed in March 1962.

Ex-CR Class 3F 0-6-0T No. 56267.
The 29 class of 1895 designed by J. McIntosh together, with the 782 class, were intended for shunting work on the Caledonian Railway system. A total of 145 engines of this type, BR Nos 56230–56376, were taken over by British Railways and were in service on the former Caledonian, Glasgow & South Western and Highland Railway sections of the Scottish Region. The majority of the class later received stovepipe chimneys. No. 56267 was built at St Rollox Works, Glasgow in 1898.

Ex-CR Class 2F 0-6-0 No. 57265. One of the numerous Drummond 'Standard Goods' engines introduced in 1883 with later additions by Lambie and McIntosh. The class, (BR Nos 57230–57473) was widely distributed throughout the Scottish Region for light freight duties and were affectionately known as 'Jumbos', No. 57265, a Neilson & Company rebuilt engine, is seen at Grangemouth in March 1956. This 0-6-0 retains its Caledonian Railway chimney – the majority of the class received the stovepipe variety whilst in British Railways service.

Ex-CR Class 3F 0-6-0 No. 57555. Designed by J. F. McIntosh for Caledonian Railway mixed traffic haulage, the 812 and 652 class 0-6-0s were introduced at St Rollox Works, Glasgow in 1899. Over 90 members of the class, BR Nos 57550–57645, were taken into British Railways ownership. No. 57555 was photographed at Glasgow Polmadie shed on 10th June 1956. One engine of the class, Caledonian Railway No. 828, survives in preservation in original blue livery.

Ex-CR Class 3F 0-6-0 No. 57667. Introduced in January 1918 by W. Pickersgill for Caledonian Railway freight haulage, the 294 and 670 class (BR Nos 57650–57691) were an enlarged version of the 812 class (above). No. 57667 is seen at Grangemouth shed in ex-works condition on 25th March 1956.

British Railways Standard Class 7MT 'Britannia' 4-6-2 No. 70052 *Firth of Tay*.
Designed at Derby, built at Crewe and introduced in 1951 under the supervision of R. A. Riddles, this class was the first of twelve Standard types for British Railways. The 2-cylinder 'Pacifics' totalled 55 engines, Nos 70000–70054. The last five, Nos 70050–70054 were allocated to Glasgow Polmadie depot post-1954, and received the names of Scottish Firths. The Scottish Region based 4-6-2s were used mainly on West Coast Main Line passenger and express freight workings. No. 70052 was recorded at Carstairs Junction in June 1958. The first engine of the class, No. 70000 has been preserved, as has No. 70013.

British Railways Standard Class 6MT 'Clan' 4-6-2 No. 72003 *Clan Fraser*.
Derby-designed and built at Crewe in 1951/2, the 'Clan' Pacifics were a lightweight version of the 'Britannia' 4-6-2 class. Only ten engines of this type were constructed, Nos 72000–72009, an order for a further 15 engines being cancelled. The Scottish allocation of the 'Clans' was shared between Glasgow Polmadie and Carlisle Kingmoor depots. No. 72003 was photographed at Polmadie in April 1956 and the engine is about to leave the depot for Glasgow Central station to work a Liverpool and Manchester express, a regular turn for these short-lived Pacifics.

British Railways Standard Class 5 4-6-0 Nos 73105 and 73109.
The first engine of this class, No. 73000, a Riddles design derived from the successful Stanier LMSR Class 5MT 4-6-0, emerged from Derby Works in April 1951. A total of 172 engines were built, Nos 73000–73171. Nos 73125–73154 were fitted with Caprotti valve gear. In 1954, Nos 73005-73009 were allocated to Perth depot and worked named expresses, the 'Saint Mungo', the 'Bon Accord' and the 'Granite City', between Glasgow Buchanan Street and Aberdeen. Doncaster-built Nos 73105 and 73109, of Glasgow Eastfield shed, were recorded leaving Edinburgh Haymarket depot in April 1956.

British Railways Standard Class 4MT 2-6-0 No. 76074.
Introduced in 1952 for mixed traffic work, 115 of these engines, Nos 76000–76114, were built at Horwich and Doncaster. The first five of the class, Nos 76000–76004 were allocated to Motherwell shed. Derived from the LMS Ivatt '43000' series 2-6-0 design of 1947, a low axle loading gave these locomotives a wide route availability. No. 76074 was photographed at Polmont in May 1957.

British Railways Standard Class 3MT 2-6-0 No. 77009.
One of a class of 20 Swindon-built engines, Nos 77000–77019, introduced in 1954 for mixed traffic duties. The main allocation of the Scottish-based locomotives post-1954 was shared between Glasgow Polmadie and Hurlford (Kilmarnock) sheds. No. 77009 is seen here at Polmadie in July 1959.

British Railways Standard Class 2MT 2-6-0 No. 78048.
A class of 65 engines, Nos 78000–78064, intro-duced at Darlington Works in 1952 and designed for mixed traffic duties. Developed from the Ivatt 2-6-0 class design for the LMS, the Scottish-based examples were mainly allocated to Edinburgh St Margaret's and Hawick depots post-1954. No. 78048, built in 1955, was recorded at Haymarket Central Junction, Edinburgh with an inspectors' saloon coach in September 1957.

British Railways Standard Class 4MT 2-6-4T No. 80025.
A class totalling 155 engines built at Brighton, Derby and Doncaster between 1951 and 1957 under the supervision of R. A. Riddles, the 2-6-4Ts Nos 80000–80154 were designed for mixed traffic work. Many of the class were based at depots in the Glasgow area and shared the Clyde coast passenger services with the ex-LMS Fairburn 2-6-4Ts from which design they were derived. No. 80025 is at Glasgow Corkerhill depot in June 1957.

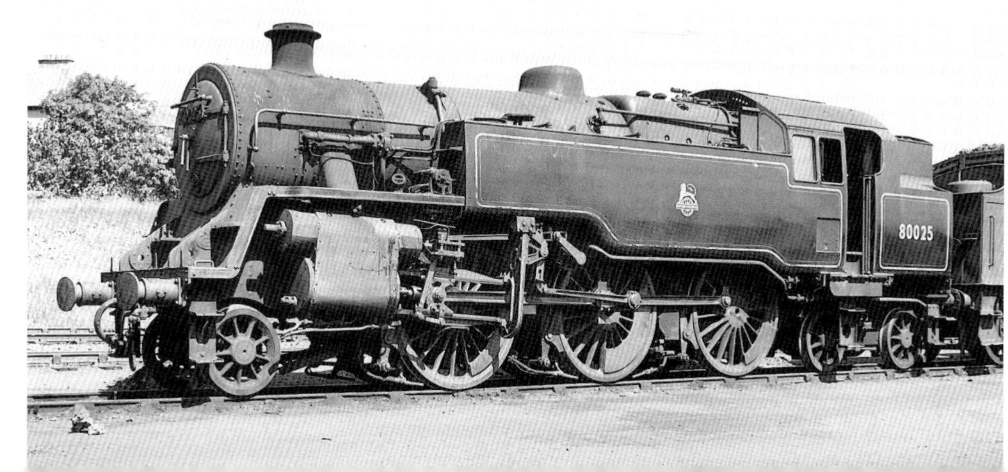

Ex-War Department 'Austerity' 2-8-0 No. 90464.

Introduced in 1943, during the Second World War, some 935 'Austerity' 2-8-0s were mass-produced for service at home and overseas and were constructed to a Ministry of Supply design by R. A. Riddles at the North British Locomotive Company, Glasgow and the Vulcan Foundry. The 'Utility' 2-8-0s eventually owned by British Railways were a large class of 733 engines, BR Nos 90000–90732. and many were based on the Scottish Region for heavy freight working. One engine of the class was named – No. 90732 *Vulcan*, the last locomotive numerically. No. 90464 is shown hard at work on a southbound evening freight train on the West Coast Main Line at Elvanfoot on 18th April 1959.

Ex-War Department 'Austerity' 2-10-0 No. 90755.
A class of 25 Crewe-designed locomotives, BR Nos 90750-90774, introduced in 1943 and built to Ministry of Supply specifications by the North British Locomotive Company, Glasgow. Used for heavy freight haulage all members of the class were based in Scotland with Motherwell and Grangemouth depots having the largest allocations. Nos 90773 and 90774 were both named *North British*. No. 90755 was photographed at Grangemouth in ex-works condition on 1st March 1959.

Scottish Steam at Work and 'On Shed'

Former North British Railway locomotive types.

For detailed descriptions of locomotive types illustrated in this Section, see Scottish Locomotive Surveys 1 and 2.

On 2nd July 1957, Class C15 4-4-2T No. 67474 leaves Craigendoran with the push-and-pull passenger train service to Arrochar and Tarbet via Helensburgh (Upper) and Garelochhead. On reaching Craigendoran Junction, the engine will propel its train northwards over the lower section of the West Highland line. The steam-hauled service between Craigendoran and Arrochar is featured on Pages 64/5.

Class C15 and C16 4-4-2Ts

Class C16 4-4-2T No. 67494 stands at the west end of Polmont shed on 21st April 1956. Regularly seen at work on the Polmont, Falkirk (Grahamstown) to Grangemouth local passenger trains, and on the Bo'ness branch, No. 67494 shared these workings with three other 4-4-2Ts based at Polmont.

Class D30/2 'Scott' 4-4-0 No. 62421 *Laird o' Monkbarns* and Class D34 'Glen' 4-4-0 No. 62483 *Glen Garry* await their turn for minor repairs on the 'dead line' sidings at Haymarket shed on 23rd September 1955.

Class D30/2 'Scott' and Class D34 'Glen' 4-4-0s

Hauling a train of vans from the Lothian Road goods yards adjoining Edinburgh Princes Street station to Slateford sidings on the western outskirts of the city, Class D30/2 'Scott' 4-4-0 No. 62419 *Meg Dods* shows evidence of overheating around the smokebox door as she passes Merchiston, on the former Caledonian Railway line between Edinburgh Princes Street and Glasgow Central on a sunny evening in June 1957.

Named after characters in Sir Walter Scott's 'Waverley' novels, the D30/2 'Scott' class 4-4-0s received hand-painted names on their driving wheel splashers. No. 62422 *Caleb Balderstone* was built for the North British Railway at Cowlairs Works, Glasgow in May 1914 and was withdrawn from British Railways service in December 1958.

D30/2 'Scott' class 4-4-0 No. 62428 *The Talisman* was built at Cowlairs Works in September 1914 and was withdrawn from service on the same date as No. 62422. Both engines were stored out of use at Bathgate depot before being sent to the scrapyards.

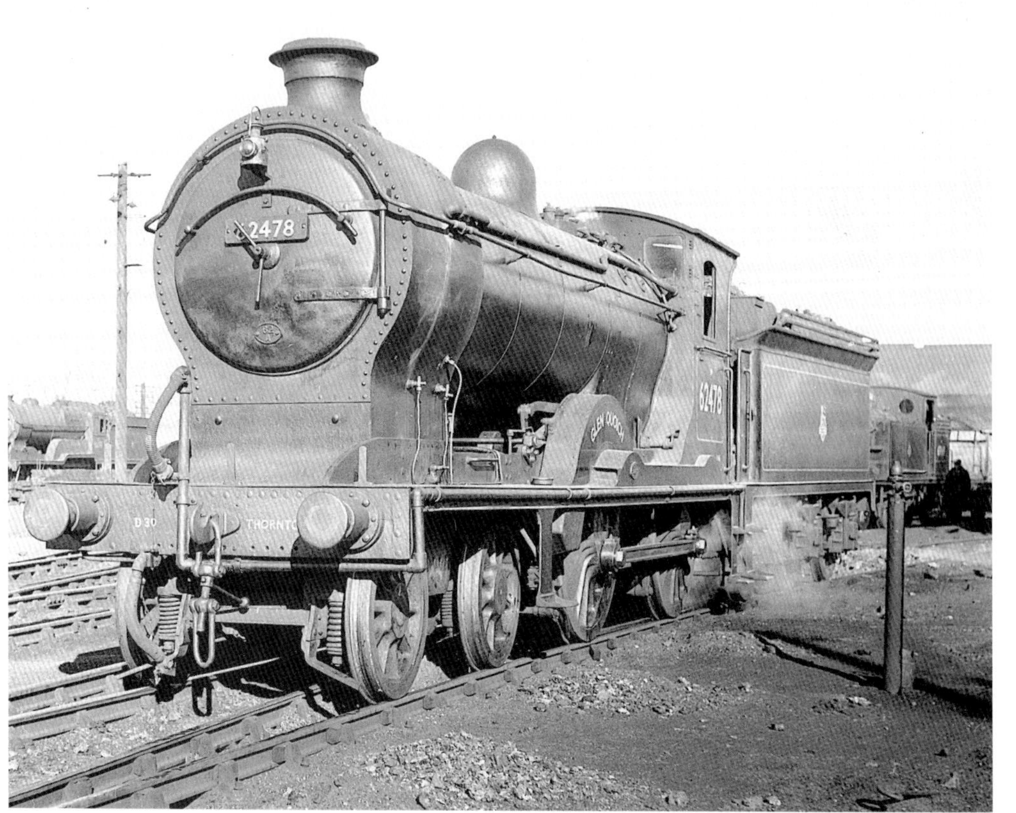

Class D34 'Glen' class 4-4-0 No. 62478 *Glen Quoich* was photographed at Haymarket shed on 15th October 1955 before working a local passenger train from Edinburgh (Waverley) to Anstruther, Fife. Fitted with a slightly extended smokebox in an attempt to overcome heating of the smokebox doors when working hard, No. 62478 was allocated to Thornton Junction depot (62A) and was withdrawn from service in December 1959 after a working life of 42 years.

Class D34 'Glen' 4-4-0 No. 62475 *Glen Beasdale* prepares to leave Haymarket shed for Edinburgh Waverley station before heading an evening passenger local train to Fife on 14th August 1958. Built at Cowlairs Works in December 1913, the former North British Railway 4-4-0 was withdrawn from service in June 1959.

Class D34 'Glen' 4-4-0 No. 62477 *Glen Dochart* in close up at Eastfield depot, Glasgow on 10th June 1956. Built at Cowlairs Works as North British Railway No. 100 in 1917, the 'Glen', based at Eastfield (65A), survived in service until its withdrawal in October 1959.

Saughton Junction, situated on the former North British Railway's Edinburgh to Glasgow main line to the west of Edinburgh, is the diverging point for the Fife, Dundee and Aberdeen lines.Class J35/4 0-6-0 No. 64534 heads a westbound freight train past the junction en route to the Glasgow area on 8th May 1956.

Class J35/4 and Class J36 0-6-0s

On 13th May 1956, Class J36 0-6-0 No. 65214 was photographed outside Kipps depot, Coatbridge in company with other North British Railway types – a Class Y9 0-4-0ST and a Class N15/1 0-6-2T.

The larger North British Railway freight engine classes were often pressed into service during busy summer holiday periods. Class J37 0-6-0 No. 64559 of Glasgow Parkhead depot (65C) rattles over the crossover between the Forth Bridge and Dalmeny station whilst working a Fife coast to Glasgow Queen Street four-coach non-corridor passenger train on 10th July 1955.

Class J37 0-6-0s

Not long after returning from works overhaul, Class J37 0-6-0 No. 64566 shunts in the yards adjoining Seafield engine depot, Leith, on 1st April 1956. Seafield, with an allocation of around 21 engines during the mid-1950s, was classified as a sub-depot of Edinburgh St Margaret's (64A) and was situated only a short distance from South Leith, a sub-shed and assembly point for eight 0-6-0T shunting engines which marshalled wagons in the Portobello area, to the east of the city.

North British Railway tank engine types

One of the former Waverley station West End 'pilot' engines, Class J83 0-6-0T No. 68478, is illustrated to show its curious unofficial livery depicting the old-style British Railways 'lion and wheel' emblem together with the initials LNER on the engine's tank sides. No. 68478 was under repair when recorded at Haymarket shed on 11th August 1958.

North British Railway Class J88 0-6-0T No. 68326, a short-wheelbase design by W. P. Reid for dockyard working, was photographed in clean condition at its home shed of Glasgow Eastfield on 24th July 1955.

Six Class Y9 0-4-0ST engines were allocated to Kipps depot, Coatbridge in the mid-1950s. Running attached to a wooden tender, No. 68116 was photographed at Kipps on 13th May 1956.

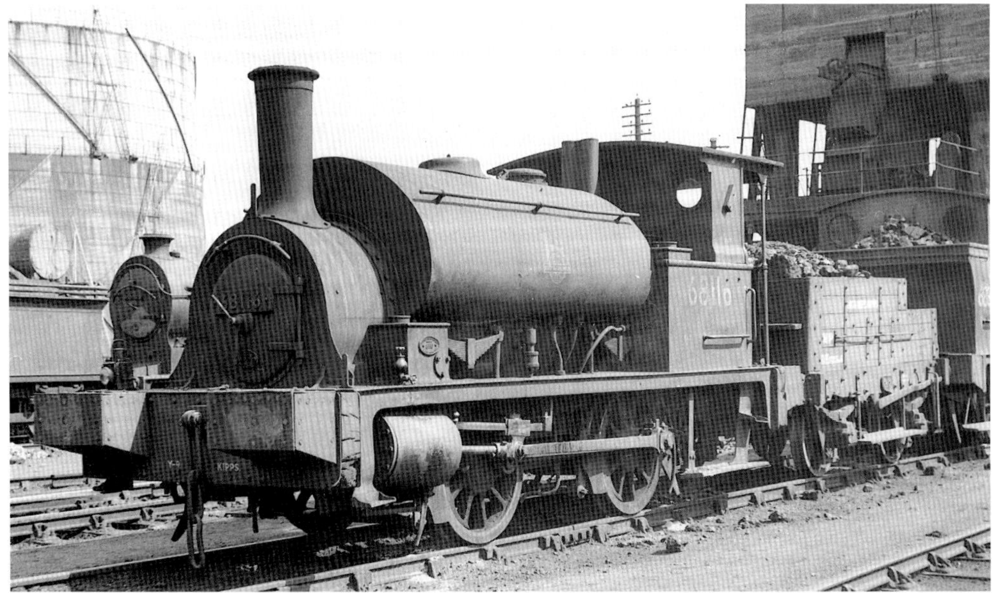

Class N15/1 0-6-2T No. 69160, a shunting engine type introduced in 1910 for the North British Railway, was recorded outside Dunfermline Upper engine depot (62C) on a summer Sunday in 1958.

Recently outshopped after major overhaul at Cowlairs Works, Glasgow, Class N15/1 0-6-2T No. 69169 is shown at Haymarket depot standing behind the corridor tender of Class A4 4-6-2 No. 60027 *Merlin* on 14th July 1955.

McIntosh Class 3P 'Caley Bogie' 4-4-0 No. 54446, one of the last survivors of the class in regular service, was recorded by the camera on station pilot duties at Carstairs Junction on 12th August 1955. No. 54446 shared station 'pilot' workings with Class 2P 0-4-4T No. 55261.

Class 3P 'Caley Bogie' 4-4-0s

After connecting with the Aberfeldy branch line train at Ballinluig, Perthshire, Class 3P 4-4-0 No. 54499 heads southwards from the junction with an Aviemore, Blair Atholl and Pitlochry to Perth stopping train on 5th July 1957.

Dalry Road depot, Edinburgh had an allocation of six former Caledonian Railway Class 2P 0-4-4Ts in the 1950s, around four of which were always well turned out for working the Edinburgh Princes Street to Leith North local branch line passenger services. Here, No. 55202 heads a four-coach passenger train for Edinburgh Princes Street near Ravelston between Craigleith and Murrayfield on 17 May 1955.

Class 2P 0-4-4Ts

Six former Caledonian Railway Class 2P 0-4-4Ts were based at Dundee Tay Bridge depot (62B) for local passenger and goods yard shunting work. Pickersgill-designed Class 2P 0-4-4T No. 55227, introduced in 1915, was photographed in the shed yards at Dundee during shunting operations on 14th April 1956.

Carstairs Junction station 'pilot'

Always maintained in immaculate condition by the shed staff at Carstairs Junction, Class 2P 0-4-4T No. 55261, the regular station 'pilot' engine, marshalls coaching stock at the station during a September weekend in 1956. No. 55261 was a post-Grouping version of the Caledonian Railway 439 class introduced by the LMSR in 1925. This engine can also be seen in action on Page 57.

Leith branch freight

The daily freight working between Leith North and Edinburgh Lothian Road goods yards is seen trundling along the former Caledonian Railway branch line behind Class 3F 0-6-0T No. 56253 in rural surroundings near Murrayfield, only a mile from the city centre. The route of the former line has become a walkway and cycle track.

Used for dockyard and timber yard shunting duties in and around the port of Grangemouth, Class 3F 0-6-0Ts Nos 56375 and 56376, the last two members of the class numerically, and identical in all respects except for different chimney lengths, stand in the shed yards at Grangemouth on 17th April 1955.

Grangemouth dockyard 0-6-0Ts

McIntosh Class 2F 0-6-0T No. 56164, a locomotive type designed for tight curve dockyard working throughout the Caledonian Railway system, was photographed in profile at Grangemouth on 31 March 1956.

Class 2F 'Jumbo' 0-6-0s

Class 2F 0-6-0 No. 57451 heads the daily 'pick-up' goods working which was returning to Carstairs Junction after shunting at intermediate stations between Elvanfoot and Thankerton on the West Coast Main Line. The 'Jumbo', with its load of various wagons, was photographed near Thankerton on 14th June 1958.

Fitted with tender cab, veteran Class 2F 0-6-0 No. 57276 is employed on station 'pilot' duties in the station yards at Oban on 22nd September 1957.

In the 1950s, Polmadie depot, Glasgow had an allocation of around 30 Class 2F 'Jumbo' 0-6-0s for shunting work in the intensive freight yards, steel works and coal mines in the Glasgow area. With original chimney and fitted with Westinghouse brake pump, No. 57465 still survived in working condition when recorded at Polmadie on 16th April 1956.

Class 3F 0-6-0s

Class 3F 0-6-0 No. 57559 leaves Craigleith station on the western outskirts of Edinburgh with an afternoon local passenger train from Leith North to Edinburgh Princes Street terminus on 22nd October 1955.

Class 3F 0-6-0 No. 57583 is panned by the camera near Crawford on the West Coast Main Line whilst heading the morning 'pick-up' goods train between Elvanfoot and Carstairs Junction, a working which is also illustrated on the opposite page.

Recently outshopped after overhaul at St Rollox Works, Glasgow, former Caledonian Railway Class 3F 0-6-0 No. 57691, numerically the last engine of the class, was photographed at Grangemouth shed on 17th April 1955.

Built in the British Railways era by Peppercorn from a Class A1/1 design for the LNER, Class A1 4-6-2 No. 60161 *North British* leaves Haymarket depot en route to Edinburgh Waverley station to work a southbound Newcastle express over the East Coast Main Line on 31st March 1956.

Main line Pacifics

Following hard on the heels of the westbound freight train illustrated on Page 154, Class A2 Pacific No. 60536 *Trimbush* accelerates the 4pm Edinburgh Waverley to Glasgow Queen Street express past Saughton Junction on 8th May 1956.

Pictured leaving Haymarket depot in tandem en route to Edinburgh Waverley station to take charge of the morning southbound expresses to Newcastle and London King's Cross, Class A3 4-6-2 No. 60083 *Sir Hugo*, based at Newcastle Heaton depot and Haymarket shed's Class A4 No. 60011 *Empire of India*, pass Haymarket Central Junction on 12th October 1958.

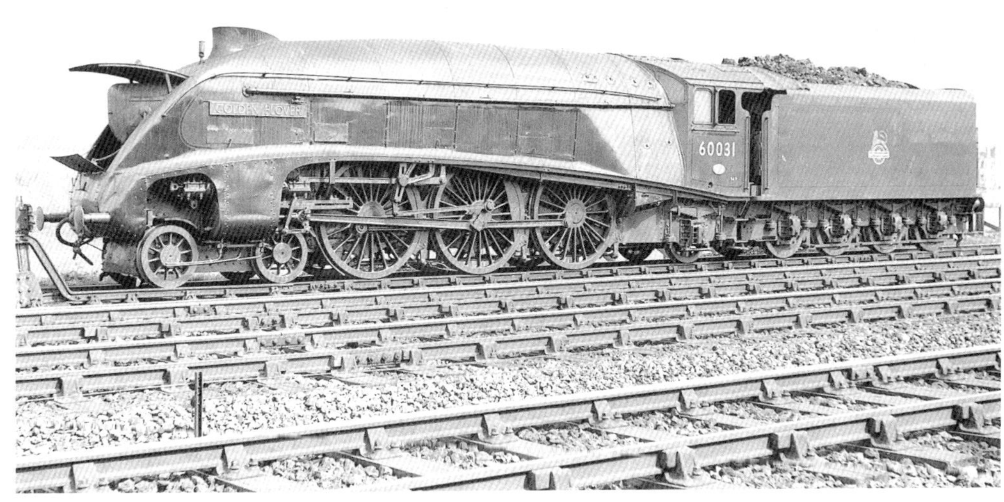

With its streamlined front and smokebox door open for inspection, one of Haymarket shed's 'top-link' Pacifics, Class A4 No. 60031 *Golden Plover* is seen outside the depot on 12th May 1958 after working the non-stop 'Elizabethan' between London King's Cross and Edinburgh Waverley.

Fitted with Kylchap blast pipe, double chimney and smoke deflectors, the unique Haymarket-based Class A3 4-6-2 No. 60097 *Humorist* climbs towards North Queensferry station with an Aberdeen to Edinburgh Waverley express on 26th May 1956.

Six-coupled mixed traffic and freight engines

Built by British Railways to a Thompson 1945 LNER design, Class B1 4-6-0 No. 61397, a later addition to a numerous class, and based at Edinburgh St Margaret's depot, heads a southbound freight train through the rock cutting at Hookhills between Inverkeithing and North Queensferry on the Aberdeen, Dundee and Fife to Edinburgh Waverley main line in the summer of 1955.

All 35 Gresley-designed Class J38 0-6-0 freight locomotives were based in Central Scotland for the haulage of heavy coal and mineral traffic. Halted at Dalmeny station before continuing north over the Forth Bridge with a Fife bound freight train, No. 65932 was recorded by the camera on 13th August 1955.

Class D49/1 'Shire' class 4-4-0 No. 62708 *Argyllshire* speeds past Dalmeny Junction signal box before braking hard for the speed restriction over the Forth Bridge with an Edinburgh Waverley to Dundee Tay Bridge passenger working on 23rd July 1955.

'Shire' and 'Director' class 4-4-0s

The D11/2 class 4-4-0s were introduced by the LNER in 1924 for passenger train haulage in Scotland. No. 62672 *Baron of Bradwardine* based at Glasgow Eastfield shed, was photographed at Jamestown Viaduct between Inverkeithing and North Queensferry with a Fife Coast to Glasgow Queen Street semi-fast passenger working in the summer of 1956.

Gresley-designed Class K2/l 2-6-0 No. 61721 leaves North Queensferry station, Fife, and heads towards the Forth Bridge crossing with a Dunfermline to Glasgow Queen Street passenger train in 1956.

Class K2/1 and Class K3/2 2-6-0s

Also designed by Gresley, Class K3/2 2-6-0 No. 61855, one of six engines of the class fitted with Great Northern Railway tenders, attacks the steep gradient between Newington and Blackford Hill stations on the Edinburgh south-side suburban line with a heavy westbound freight train on 29th June 1956.

Six-coupled mixed traffic and suburban passenger locomotives

Class K4 2-6-0 No. 61995 *Cameron of Lochiel*, fitted with a straight steampipe above the cylinders on one side and an angled steampipe on the other side, leaves a smoke-screen behind whilst working an Edinburgh to Glasgow Cadder westbound freight train near Polmont on 25th May 1957.

On 14th August 1958, clean green-liveried Class V2 2-6-2 No. 60819 based at Aberdeen Ferryhill depot (61B), backs down towards Haymarket shed after being coaled and watered. Later, the V2 will return north with an evening passenger train working to Dundee and Aberdeen.

From the footplate, the fireman of Class V1 No. 67615 admires the view westwards along the estuary of the River Forth towards Rosyth Naval Dockyard as the 2-6-2T climbs the gradient to the Forth Bridge with a Dunfermline to Edinburgh Waverley local passenger train on a summer's afternoon in July 1958.

One of the last Scottish based Class 4P 3-cylinder Compound 4-4-0s to remain in service, No. 40920, and one of five examples of the class based at Ayr depot, (67C), shows its classic Midland Railway lines to advantage in this broadside shot taken at Stranraer on 3rd August 1956. The 4-4-0 was about to leave the Town station with a passenger train for Ayr and Glasgow St Enoch, a photograph of which appears on Page 49.

Scottish Midland Compound 4-4-0s

On a May morning in 1955, an Edinburgh Princes Street to Stirling passenger train passes Haymarket on the former Caledonian Railway spur line to Haymarket West Junction, behind Class 4P Compound 4-4-0 No. 40938, based at Perth depot (63A).

On the former Glasgow & South Western Railway line between Irvine and Ayr, Class 5MT 2-6-0 No. 42740 heads a freight train for Ayr through the well-kept Barassie station north of Troon on a summer's day in 1961.

South-West Scotland-based 'Crab' 2-6-0s

At St Rollox depot, Glasgow, Class 5MT 2-6-0 No. 42915, fresh from overhaul at the nearby locomotive works, awaits return to its home shed at Dumfries on 17th August 1958. It is doubtful if the paint-shop foreman at St Rollox Works would have approved of the application of the engine's cab-side number in such an off-centre position!

Shutting off steam and braking hard on the approaches to the Forth Bridge speed restriction, Dundee Tay Bridge depot-based Stanier Class 5MT4-6-0 No. 45384, with burnished smokebox handles and hinges, leans to the curve near Dalmeny Junction with an Edinburgh Waverley to Dundee passenger train working on 13th August 1955.

Scottish-based Stanier Class 5MT 4-6-0s

Heading into the night, Stanier Class 5MT 4-6-0 No. 44668, a Carlisle Kingmoor engine, makes steady progress along a straight section of the West Coast Main Line near Thankerton with a southbound fitted van train from Glasgow on 10th April 1960.

With a typical 'Jubilee' class roar from her exhaust when working hard, Class 6P 'Jubilee' 4-6-0 No. 45718 *Dreadnought*, based at Carlisle Kingmoor shed, was captured in action by the camera powering southwards near Symington on the West Coast Main Line with a Glasgow Central to Liverpool and Manchester express in July 1958.

'Jubilee' class 4-6-0s at work

In quieter mood, Class 6P 'Jubilee' 4-6-0 No. 45661 *Vernon* leaves Elvanfoot station with an afternoon stopping passenger train from Carlisle to Glasgow Central on 11th April 1960.

A low-angle shot taken into the early morning sun highlights the powerful lines of Class 7P 'Royal Scot' 4-6-0 No. 46104 *Scottish Borderer* as she slowly approaches Carstairs Junction station with the overnight Manchester and Liverpool to Glasgow Central sleeping-car express on 13th September 1958.

Scottish 'Royal Scot' class 4-6-0s at work

A hot day accounts for the lack of exhaust from Class 7P 'Royal Scot' 4-6-0 No. 46105 *Cameron Highlander*, one of five engines of the class allocated to Glasgow Polmadie depot (66A) in the 1950s, as she tackles the last mile of the climb to Beattock Summit near Harthope Viaduct with a 'relief' northbound passenger working from the Midlands in June 1958.

‘Princess Royal’ Class 8P 4-6-2 No. 46203 *Princess Margaret Rose*, now preserved along with No. 6201 *Princess Elizabeth*, waits to leave Polmadie depot, Glasgow before backing down to Central station to work an evening express to London Euston over the West Coast Main Line in May 1958.

‘Princess Royal’ Pacifics

The valve motion and front-end detail of record-breaking Stanier ‘Pacific’ No. 46201 *Princess Elizabeth*, a regular performer on West Coast and Birmingham New Street to Glasgow Central express workings in the 1950s and 1960s. This famous engine is also illustrated at work on Pages 43 and 56.

A 'filling-in' turn between main line duties for Class 8P 'Duchess' 4-6-2 No. 46220 *Coronation* as she heads the lightweight 1.30pm all-stations Edinburgh Princes Street to Glasgow Central passenger train through Slateford Junction on 29th September 1956. As a reminder of its former stream-lined days, No. 46220 still retains its sloping smokebox front in contrast to the cylindrical version fitted to the 'Pacific' in December 1955, a photograph of which is reproduced on Page 18.

Glasgow Polmadie-based 'Duchess' Pacifics

En route to London Euston from Glasgow Central via Carlisle, the up 'Royal Scot' hauled by a well-polished Class 8P 'Duchess' 4-6-2 No. 46222 *Queen Mary,* takes water at Strawfrank troughs to the south of Carstairs Junction in the summer of 1958.

Carrying a blue-backed nameplate, Class 8P 'Duchess' 4-6-2 No. 46231 *Duchess of Atholl* speeds the up 'Royal Scot' between Crawford and Elvanfoot on the West Coast Main Line in July 1960. No. 46231 was one of the first three engines of the class to be scrapped in 1962.

Between Harthope Viaduct and Beattock Summit, a well-known location for generations of railway photographers, Class 8P 'Duchess' 4-6-2 No. 46232 *Duchess of Montrose*, built at Crewe Works on 1st July 1938, heads north with a lightweight London Euston to Perth express on a bright summer's evening in July 1960. First allocated to Glasgow Polmadie depot in 1940, No. 46232 remained a Glasgow-based engine until its withdrawal from service in December 1962.

British Railways Class 7 'Britannia' 4-6-2 No. 70047, the only member of the class never to receive a name, heads southwards through Elvanfoot with a late afternoon express from Glasgow Central to London Euston on 7th July 1960.

'Britannia' and 'Clan' class 4-6-2s at work and 'on shed'

Between main line duties on the West Coast route, Class 7 'Britannia' 4-6-2 No. 70052 *Firth of Tay* was recorded 'on shed' at Polmadie depot, Glasgow on 24th July 1955.

The first engine of a British Railways Standard class which totalled only ten locomotives, Class 6 'Clan' 4-6-2 No. 72000 *Clan Buchanan* hurries southwards near Crawford with the morning Glasgow Central to Manchester and Liverpool express on 11th April 1960.

Designed at Derby and introduced by British Railways in 1952, Class 6 'Clan' 4-6-2 No. 72002 *Clan Campbell* was recorded at Perth depot on 2nd June 1956 after its arrival at Perth General station with a parcels train from Glasgow Buchanan Street, an interesting comparison with the 'Britannia' illustrated opposite.

Designed at Doncaster and a class which was first introduced by British Railways in 1951 for mixed traffic duties, Class 5 4-6-0 No. 73076 heads up Beattock Bank with an afternoon Carlisle to Carstairs Junction and Glasgow Central stopping passenger train on 15th August 1955.

British Railways Standard Class 5 4-6-0s in action

Standard Class 5 4-6-0 No. 73109 piloting Class B1 4-6-0 No. 61219 passes Haymarket Central Junction with a Glasgow Queen Street to Edinburgh Waverley express passenger train on 1st September 1957.

British Railways Standard 2-6-0 on freight

After a permanent way signal check to the north of Elvanfoot on the West Coast Main Line, British Railways Standard Class 4 2-6-0 No. 76070 picks up speed with a Carlisle Kingmoor yard to Motherwell freight working in July 1960.

British Railways Standard 2-6-4T on local passenger

British Railways Standard Class 4 2-6-4T No. 80109 sporting a red-backed front number plate and carrying a former Caledonian Railway style route indicator, heads away from a station stop at Slateford with the 1.30pm Edinburgh Princes Street to Glasgow Central passenger train via Mid Calder and Shotts on 31st August 1957.

Miscellany

Saved from the scrap heap

Class D49/1 'Shire' 4-4-0 No. 62712 *Morayshire*, built at Darlington in February 1928 as LNER No. 246, was withdrawn from Hawick shed in July 1961. Renumbered 2712 in 1946 and No. 62712 by British Railways in 1948, the engine was photographed in derelict condition, minus rods and motion at Edinburgh, Dalry Road shed prior to its journey north for restoration at Inverurie Works, Aberdeenshire. No. 246, preserved in its original LNER apple green livery, is now owned by the Royal Museum of Scotland and is at present in the care of the Scottish Railway Preservation Society at Bo'ness.

Built at St Rollox Works, Glasgow in November 1907, Class 2P 0-4-4T No. 55189 (CR No. 419) of the 439 class is a variant of the numerous Caledonian Railway locomotives of this type which appeared between 1884 and 1925. The engine was recorded at Glasgow Polmadie depot on 16th April 1956. Withdrawn from Carstairs Junction depot in 1962 the locomotive was acquired for preservation and restored to its original blue livery at Cowlairs Works, Glasgow. CR No. 419 is also in the care of the Scottish Railway Preservation Society at Bo'ness and is maintained in working order.

Recorded at Glasgow Corkerhill depot in the summer of 1956, ex-Caledonian Railway 812 class 0-6-0 No. 57566 was built at St Rollox Works, Glasgow, in August 1899. Withdrawn from British Railways service at Ardrossan in 1963, the engine was also acquired for preservation and has been restored to Caledonian Railway blue livery as No. 828. The engine, previously housed in the Glasgow Museum of Transport is at present at the Strathspey Railway, Aviemore, Inverness-shire.

Built by the LNER at Doncaster Works as No. 4488 in June 1937 and withdrawn from regular service in June 1967 from Aberdeen Ferryhill depot, Gresley Class A4 4-6-2 No. 60009 *Union of South Africa* was the only Scottish-based preserved 'Pacific' locomotive and was originally located at the now-closed Lochty Private Railway near Anstruther, Fife. To coincide with the Forth Bridge centenary celebrations in 1990, the streamlined engine was temporarily renamed *Osprey*. Photographed passing through Princes Street Gardens, Edinburgh, the A4 gets into her stride with a north-bound special train in August 1980. Allocated to Edinburgh Haymarket depot in the 1950s, No. 60009 was a regular performer on the non-stop 'Capitals Limited' and the 'Elizabethan' East Coast Main Line expresses between Edinburgh Waverley and London King's Cross.

No. 9 at work

The Gresley Class A4 4-6-2 No. 60009 has been used on numerous charter trains and special rail trips since its preservation. It is seen here, accelerating the 'Aberdonian' past Haymarket West Junction to the west of Edinburgh with a Kirkcaldy Lion's Club special train in the 1970s.

Designed by Matthew Holmes, Class J36 0-6-0 goods engine, formerly North British Railway C class, No.65243 was built by Neilson & Company, Glasgow in December 1891. Rebuilt in December 1915, the engine, together with 24 other members of the class, was sent to France with the Railway Operating Department on war-time service to work supply trains to the Western Front and on return received the name *Maude* after a military general of the Coldstream Guards. Renumbered 9673 by the LNER in 1924, the 0-6-0 became No. 5243 in 1946 and finally British Railways No. 65243 in June 1948. Allocated to Edinburgh Haymarket depot for most of its career, *Maude* was withdrawn from Bathgate shed in July 1966 and now survives in preservation at Bo'ness in NBR goods engine livery. The engine is seen here at Haymarket shed on 6th May 1956.

North British Railway
0-6-0 *Maude*

On one of its early appearances on the main line after preservation, the veteran ex-North British Railway 0-6-0 No. 673 *Maude* emerges from the short tunnel to the south of Inverkeithing station on the Aberdeen to Edinburgh Waverley route. The engine raises the echoes on the 1 in 70 gradient towards North Queensferry on the approaches to the Forth Bridge with a Scottish Railway Preservation Society special train carrying a 'Fife Coast Express' headboard on 4th May 1980.

In conjunction with the Scottish Industries Exhibition held at the Kelvin Hall, Glasgow in 1959, passenger trains were run to the exhibition centre from many parts of Scotland using the four restored Scottish pre-Grouping locomotives. Caledonian Railway 4-2-2 No. 123 piloting Great North of Scotland Railway 4-4-0 No. 49 *Gordon Highlander* is shown after arrival at Kelvin Hall station with one of the special trains in September of that year.

Scottish pre-Grouping locomotives restored

In company with a variety of ex-Caledonian Railway 0-6-0s, Great North of Scotland Railway 4-4-0 No. 49 *Gordon Highlander* catches the afternoon sun at Dumfries depot. The preserved engine, resplendent in its original green livery, was being prepared at the shed to work the final leg of a return journey from Dumfries to Glasgow St Enoch via Kilmarnock with a Stephenson Locomotive Society special train on 13th June 1959, its first main line outing since its restoration.

Designed by Sir Henry Fowler, the first 50 engines of the LMSR 'Royal Scot' class 4-6-0s were built in 1927 by the North British Locomotive Company at their Hyde Park and Queen's Park works in Glasgow. Afurther 20 engines were built at Derby in 1930. This photograph, taken at Glasgow Central station, shows the leading driving wheel splasher and nameplate *The East Lancashire Regiment* surmounted by the cast brass regimental crest. No. 6135, one of the Hyde Park batch of engines built in September 1927 and originally named *Samson*, was renamed by the LMS in 1936. Rebuilt with taper boiler and double chimney by Sir William Stanier in January 1947, No. 46135 was scrapped in December 1962.

Locomotive nameplates

The nameplate of the LMSR Stanier Pacific No. 46200 *The Princess Royal,* delivered to traffic from Crewe Works in June 1933. The first engine of a class of 13 locomotives including the 'Turbomotive', No. 46200, was briefly allocated to Glasgow Polmadie depot in 1951. In British Railways days, the 'Princess Royal' class carried a variety of liveries including London & North Western Railway lined black, blue and later, green. Four members of the class received crimson lake livery in 1958 including No. 46200. A regular performer on Anglo-Scottish expresses over the West Coast Main Line, *The Princess Royal* was the last engine of the class to be withdrawn from service at Carlisle Kingmoor depot in November 1962. Two engines of the class survive in preservation, Nos 6201 and 46203, and are both maintained in working order.

The nameplate and company crest of the restored Great North of Scotland Railway 4-4-0 No. 49 *Gordon Highlander*. Classified as D40 by the LNER and known locally in the Aberdeen area as 'The Sojer', the 4-4-0 was built by the North British Locomotive Company Glasgow in 1920 to a design by T. E. Heywood. No. 49 became LNER No. 6849 in 1924, No. 2277 in 1946, and finally British Railways No. 62277 at Nationalisation in 1948. *Gordon Highlander*, the last ex-GNSR locomotive at work on British Railways, was withdrawn from Keith shed in 1958 and subsequently underwent restoration at Inverurie Works to its original condition. The 4-4-0 worked enthusiasts' special trains before its retirement as a static exhibit at the former Coplawhill tramway works, Glasgow. This was later closed, and the engine has since been rehoused at the Museum of Transport at the Kelvin Hall.

Front end detail of Gresley Class K4 2-6-0 No. 61995 showing the blue-backed brass nameplate *Cameron of Lochiel*. The third member of a class comprising five locomotives, No. 61995, was built at Darlington as LNER No. 3443 and was put into service on the West Highland line in 1937/8.

No. 61995 was withdrawn in October 1961 and one of its nameplates was presented to the Cameron of Lochiel at Achnacarry Castle near Spean Bridge to commemorate his chairmanship of the British Railways Board (Scotland).

During visits to Scotland's capital city by royalty, the Royal Train was normally routed over the West Coast Main Line via the east spur line at Carstairs Junction. In charge of the special train on the occasion of the Queen's visit to Edinburgh on 3rd July 1956 were Carlisle Kingmoor shed's Stanier Class 5MT 4-6-0s, with No. 45126 piloting No. 45364. The leading locomotive carries four white headlamps denoting a Royal Train and the special working is seen travelling at reduced speed, half a mile from its destination at Princes Street station terminus, Edinburgh.

Royal Trains

On a Royal visit to Scotland four years later, on 22nd June 1960, the Royal Train was worked by the customary Carlisle Kingmoor shed's Stanier Class 5MT 4-6-0s, in tandem. Here, the specially prepared engines, Nos 44676 and 44993, reverse out of Dalry Road depot, Edinburgh, before coupling up to the Royal Train at Princes Street station for the return overnight journey to London Euston. All evidence of this once-busy railway centre has disappeared and the area now forms part of the city's Western Approach road system.

A non-stop steam-hauled special train run from London King's Cross to Edinburgh Waverley was organised by the Stephenson Locomotive Society and the Railway Correspondence & Travel Society on the weekend of 2nd and 3rd June 1962. The 'Aberdeen Flyer' left King's Cross at 8 am hauled by the famous Class A4 4-6-2 No. 60022 *Mallard* in immaculate British Railways Brunswick green livery and with burnished buffers. The train is pictured at the London terminus just prior to departure. From Edinburgh, the train continued to Aberdeen behind Class A4 4-6-2 No. 60004 *William Whitelaw*. Leaving Aberdeen southbound at 11 pm, Class 8P 4-6-2 'Princess Royal' No. 46201 *Princess Elizabeth* was in charge of the special train on the return working via the West Coast route as far as Carlisle where sister engine No. 46200 *The Princess Royal* took over for the remainder of the overnight journey, arriving at London Euston at 11 am the next day.

Mallard – Scotland bound

At London King's Cross station prior to its northbound journey with the 'Aberdeen Flyer' on 2nd June 1962, a BBC television cameraman smartly dressed to match the occasion, films the commemorative plaque on *Mallard's* streamlined casing, which records the Class A4's epic run in July 1938 when a world speed record for steam traction of 126 mph was achieved down Stoke Bank between Grantham and Peterborough on the East Coast Main Line.

Built by the London & North Western Railway in 1908 for the 'American Special' boat trains which ran between London Euston and Liverpool Riverside, a former first class dining car was still in use by the British Railways Civil Engineering Department when recorded in grey livery in the goods yards at Perth in September 1965. Originally numbered 311, the coach was renumbered 7558 and was finally withdrawn from LMS service in 1938. The elegant twelve-wheeled vehicle became ARP Lecture Car No. 7 during the Second World War and later carried the service number DM198532.

Railway relics

A station platform gas lit cast-iron lamp-post at Kinghorn station, Fife, a relic of the old Edinburgh & Northern Railway which, although administered from the capital city, ran its railway services in Fife on the northern side of the River Forth. Sir Thomas Bouch, who resided at Oxford Terrace, Edinburgh became general manager and chief civil engineer of the railway in 1849 at the age of 27. The photograph was taken in July 1955.

Railway architecture

Reaching an altitude of 1,498 feet above sea-level, the 7¼-mile long branch line from Elvanfoot, Lanarkshire to the villages of Leadhills and Wanlockhead was the highest standard gauge line in the country. Built by the Caledonian Railway in 1901/2, the line served the two isolated communities in the Lowther Hills. During its operation the light railway was worked by a Caledonian Railway 0-4-4T and later by an ex-London & North Western Railway 0-4-0T steam rail motor. The unprofitable and remote line was closed completely in 1939. The view taken in April 1959, looks south from above the derelict Leadhills Viaduct and shows the trackbed of the abandoned single line railway as it follows the hillside contours before joining the West Coast Main Line at Elvanfoot. The viaduct was demolished on 23rd December 1991.

A general view looking north of the attractive Barassie station, Ayrshire, the railway junction for Kilmarnock on the former Glasgow & South Western Railway main line between Ayr and Ardrossan. The immaculate trim lawns and ornamental flower beds were a typical feature of many of the Clyde coast resort stations in the 1950s. The line to Kilmarnock branches off sharply to the right of the picture.

The Gretna Memorial at Rosebank Cemetery, Leith, was erected as a tribute to the 227 soldiers of the 1/7th Battalion of the Royal Scots regiment bound for Gallipoli who died in the tragic railway accident at Quintinshill on the Caledonian Railway, West Coast Main Line between Kirkpatrick and Gretna, Dumfriesshire on 22nd May 1915. The memorial, consisting of a Celtic cross in red Peterhead granite flanked by ten bronze tablets, was unveiled on 16th May 1916 by Lord Rosebery, the Hon. Col. of the regiment.

In memoriam

The carved memorial stone at the Dean Cemetery, Edinburgh, dedicated to Sir Thomas Bouch, the son of a Cumberland sea captain and the railway builder and engineer who died at Moffat, Dumfriesshire in 1880, ten months after the collapse of his ill-fated Tay railway bridge which spanned the river estuary between Dundee and Wormit. The inscription reads simply:

Sir Thomas Bouch
Civil Engineer
Born 25 Feb. 1822
Died 30 Oct. 1880